Advances in Biochemistry and Biotechnology

The Editor

Dr. Chiranjib Chakraborty, Head, Department of Biotechnology, Institute of Applied Medicines & Research, is well known for his notable contribution of different technology development in biological field. A Ph.D. from Vidyasagar University (West Bengal) and Marine Aquarium and Research Center (Z.S.I) did his post - doctoral research in USA. He is a recipient of several medals and awards including Snehalata Banerjee Gold medal from the Academy of Environmental Biology for the publication of best paper in 1998. Dr. Chakraborty started his professional career as Lecturer in the department of Industrial Fish and Fisherey, Burdwan Raj College, West Bengal and also worked as R&D Manager (Biotechnology) Glenmark Lab, Mumbai, and Research Scientist in Biotechnology at Macleod Pharmaceuticals, Mumbai. He visited several institutes like IIT, Kharagpur; IIT, New Delhi; All India Institute of Medical Science, New Delhi; Department of Biotechnology; Jadavpur University, Calcutta; ERI, Chennai and Lee and Nee Software export Ltd, Calcutta for exchange of his scientific ideas. He has a good number of original research papers, review articles and four books published both within and outside India. His unstinted efforts in research have enriched seed production biotechnology of prawns; production technology development & stability of recombinant pharmaceutical proteins and diagnostics technology development through computational biology like protein microarray.

Advances in Biochemistry and Biotechnology

— *Volume 1* —

Edited by
Chiranjib Chakraborty
Head, Department of Biotechnology, Institute of Applied Medicines & Research
9th Mile Stone, Duhai, Delhi-Meerut Road, Ghaziabad.
(Previously Manager, Biotechnology (R&D), Glenmark Laboratories, Mumbai;
Research Scientist, Biotechnology, Mucleods Pharmaceuticals Ltd., Mumbai;
Lecturer, Deaprtment of Industrial Fish and Fishery, Burdwan Raj College, Burdwan, West Bengal, India)

2027
Daya Publishing House®
A Division of
Astral International Pvt. Ltd.
New Delhi - 110 002

First Published, 2005

Reprinted 2027

ISBN:978-81-7035-362-1 (Hardbound)

Published by : **Daya Publishing House®**
A Division of
Astral International Pvt. Ltd.
– ISO 9001:2008 Certified Company –
4760-61/23, Ansari Road, Darya Ganj
New Delhi-110 002
Ph. 011-43549197, 23278134
E-mail: info@astralint.com
Website: www.astralint.com

Laser Typesetting : **Classic Computer Services**
Delhi - 110 035

Digitally Printed at : **Replika Press Pvt. Ltd.**

PRINTED IN INDIA

Preface

With all optimistic view, we admit that we know very little of the whole nature and its governing. We are still trying to understand nature. We are still trying to understand. We are trying to decipher the codes used for storing various information with the advanced technology of Biological sciences i.e. Biotechnology and Biochemistry, which have developed since last 50 years.

This book has been designed to incorporate contributors of several Indian scientists of admirable skill and standing in India who are participated into the advancement of this subject.

I would like to thank all the contributors for their contributions into this volume.

Views expressed in the contributions are exclusively those of the authors and responsibilities do not lie on the editors and the publisher.

Last but not the least; I take the privilege to admiring the Managing Director of Daya Publishing House, Mr. Anil. K. Mittal thankfully for completing this publication in a hot hurry.

Dr. Chiranjib Chakraborty

Contents

List of Contributors

A. Ghosh

Mosquito Research Unit, Department of Zoology, Burdwan University, West Bengal, India

Akash Verma

School of Biotechnology, Guru Ghasidas University, Bilaspur, Koni, Chhattisgarh

Anirban Banerjee

Department of Pharmaceutical Technology, National Institute of Pharmaceutical Education and Research, Sector 67. S.A.S. Nagar–160 062, Punjab

Arun Sharma

Head, Food Technology Division, B.A.R.C., Mumbai–400 085, Tel: 91-22-25505180/25505314, Fax: 91-22-25505151/25519613, Email: ksarun@apsara.barc.ernet.in

Arup Tewary

Aquaculture Research Unit, Department of Zoology, Vidyasagar University, Midnapore–721 102, West Bengal, India

Bidhan C. Patra

Aquaculture Research Unit, Department of Zoology, Vidyasagar University, Midnapore–721 102, West Bengal, India, Telephone: Office: (91) 03222-62297, Residence: (91) 03222-66035, Fax No.: (91) 03222-62329, e-mail: bcpatra@yahoo.com

Biplab Sarkar

Central Institute of Fresh Water Aquaculture, Kausalyaganga, Bhubaneswar, Orissa

Brajesh Barse

Department of Pharmaceutical Technology, National Institute of Pharmaceutical Education and Research, Sector 67. S.A.S. Nagar–160 062, Punjab

Chiranjib Chakraborty

Head, Department of Biotechnology, Institute of Applied Medicines & Research, 9th Mile Stone, Duhai, Delhi-Meerut Road, Ghaziabad

Debabrata Mukherjee

Lab. of National Fellow, Central Inland Fisheries Research Institute, C.G.O. Complex, Salt Lake, Kolkata- 64

G. Chandra

Mosquito Research Unit, Department of Zoology, Burdwan University, West Bengal, India

G.C. Mishra

Director, National Centre for Cell Science, NCCS Complex, Ganeshkhind, Pune – 411 007, Maharashtra, Tel: 020-5491065, Fax: 020-5697756, E-mail: gcmishra@nccs.res.in

Garima Sharma

Centre for Biomedical Engineering, Indian Institute of Technology Delhi, Hauz Khas, New Delhi–110 016

J.K. Deb

Department of Biochemical Engineering and Biotechnology, I.I.T., Delhi, New Delhi–110 016

M.A. Haniffa

Centre for Aquaculture Research and Extension, St. Xavier's College, Palayamkottai, Tirunelveli–627 002, Tel: +91 462 2560670: Fax: +91 462 2560670; E-mail: haniffacare@hotmail.com

Mani Shankar Bhattacharyya

Department of Pharmaceutical Technology, National Institute of Pharmaceutical Education and Research, Sector 67. S.A.S. Nagar–160 062, Punjab

Pankaj Soni

Department of Pharmaceutical Technology, National Institute of Pharmaceutical Education and Research, Sector 67. S.A.S. Nagar–160 062, Punjab

Prashant Mishra

Department of Biochemical Engineering and Biotechnology, Indian Institute of Technology Delhi, Hauz Khas, New Delhi–110 016

Prveen Kaul

Department of Pharmaceutical Technology, National Institute of Pharmaceutical Education and Research, Sector 67. S.A.S. Nagar – 160 062, Punjab

R.K. Saxena

Centre for Biomedical Engineering, Indian Institute of Technology Delhi, Hauz Khas, New Delhi–110 016

Ragini Gothalwal

Head of the Department, School of Biotechnology, Guru Ghasidas University, Bilaspur, Koni, Chhattisgarh, E-mail: ragini_gothalwal@hotmail.com

Rohit Sharma

Department of Pharmaceutical Technology, National Institute of Pharmaceutical Education and Research, Sector 67. S.A.S. Nagar–160 062, Punjab

S. Adhikari

Central Institute of Fresh Water Aquaculture, Kausalyaganga, Bhubaneswar, Orissa

S. Ayyappan

Deputy Director General, (Fisheries), ICAR, Krishi Anusandhan Bhaban, PUSA, New Delhi

S.A. Bapat

National Centre for Cell Science, NCCS Complex, Ganeshkhind, Pune 411 007, Maharashtra

Satyendra Gautam

Food Technology Division, Bhabha Atomic Research Centre, Mumbai–400 085

Sawraj Singh

Department of Pharmaceutical Technology, National Institute of Pharmaceutical Education and Research, Sector 67. S.A.S. Nagar–160 062, Punjab

Shyam Sundar Nandi

School of Biotechnology, Guru Ghasidas University, Bilaspur, Koni, Chhattisgarh, E-mail: nandibiotech@yahoo.co.in

Surajit Sinha

National Centre for Cell Science, NCCS Complex, Ganeshkhind, Pune 411 007, Maharashtra

T.K.Maiti

Department of Biotechnology, I.I.T.-Kharagpur (W.B), E.Mail- tkmaiti@hijli.iitkgp.ernet.in

U.C. Banerjee

Department of Pharmaceutical Technology, National Institute of Pharmaceutical Education and Research, Sector 67. S.A.S. Nagar–160 062, Punjab, Phone: 91-172-2214682-85 Extn. 2142, Fax 91-172-22114692, E-mail: ucbanerjee@niper.ac.in

Advances in Biochemistry and Biotechnology Volume 1 *Pages 1-9*
Edited by: **Dr. Chiranjib Chakraborty**
Published by: **DAYA PUBLISHING HOUSE**

From Editor Desk

Current Changing Scenario of Biotechnology and Biochemistry

Chiranjib Chakraborty
Head, Department of Biotechnology, Institute of Applied Medicines & Research, 9th Mile Stone, Duhai, Delhi-Meerut Road, Ghaziabad

Biotechnology and biochemistry has involved with many parts of the chemical industry, agriculture, and medicine. These two subjects have demarcation between basic and applied research, and new discoveries and innovations, in most cases, can find direct application. Innovations, techniques, and tools that have emerged and revolutionized modern biotechnology include genetic engineering, cell fusion technology, bioprocess technologies, and structure-based molecular designs including drug development, drug targeting, and drug delivery systems.

After 20 years of growing pains, the biotechnology sector is coming into its own as a mature industry, as highlighted in the article on Nature Reviews Drug Discovery (1). This is reflected in the number of companies that have made lucrative licensing deals with toptier pharmaceutical companies, as well as in the growing number of biotech companies that are on their way to becoming fully integrated drug development enterprises. With several leading biotech companies posting billion dollar earnings in 2001, there is little doubt that the major players in this industry have discovered a formula for success.

Leading biotech companies can now successfully compete with big pharmaceutical companies for the rights to late-stage compounds, and mid-size biotechs are using their considerable cash reserves to strike deals with smaller companies in attempts to become more integrated, product-focused organizations (2). Smaller biotechs have benefited by having the option of negotiating more equitable deals with other biotechs, who might have more at stake than big pharma companies and therefore be more committed to the success of a particular product. In 2001, there were more than 1,500 biotech companies in Europe and approximately 1,300 in the United States, but 2001 revenues for European biotechs were less than one-third that of US revenues (US $8 versus US $25 billion, respectively)(3). Most of the profitability within this sector came from those companies who have expanded their discovery and development capabilities over the years–in part, by making deals with or acquiring smaller, specialized biotech companies. Looking ahead, further intra-industry deal making and eventual consolidation within the biotech industry is inevitable if biotech companies are to survive to profitability.

Indian Biotechnology and Biochemistry Advancement

In the 1980s the Government of India considered the need for creating a separate institutional framework to strengthen biology and biotechnology research in the country. Scientific agencies supporting research in modern biology included: Council of Scientific and Industrial Research (CSIR), Indian Council of Agricultural Research (ICAR), Indian Council of Medical Research (ICMR), Department of Science and Technology, and University Grants Commission. Biotechnology was given an important boost in 1982 with the establishment of the National Biotechnology Board. Its priorities were human resource development, creation of infrastructure facilities, and supporting research and development (R and D) in specific areas. The success and impact of the National Biotechnology Board prompted the Government to establish a separate Department of Biotechnology (DBT) in February 1986. There have been major accomplishments in areas of basic research in agriculture, health, environment, human resource development, industry, safety, and ethical issues (4).

India led through the work of G.N. Ramachandran, in which he elucidated the triple helical structure of collagen. The Ramachandran plot has proven to be fundamental in solving the protein structure. Areas of biosystematics using molecular approaches, mathematical modeling, and genetics including genome sequencing for human beings, animals, and plants, will continue to have priority as we move into the next century. The tremendous impact of genome sequencing is increasingly evident in many fields. As an increasing number of new genes are discovered, short, unique, expressed sequenced tags segments are used as signatures for gene identification. The power of high throughput sequencing, together with rapidly accumulating sequenced data, are opening new avenues in biosciences. In the plant genome area, the sequencing of *Arabidopsis* and rice genome will soon be completed and cataloging and mapping of all the genes will be done.

There have been major achievements in basic bioscience in the last decade or so in India, where we have expertise in practically all areas of modern biology. The institutions under the CSIR, ICMR, ICAR, DST, and DBT have established a large number of facilities where most advanced research work in biosciences is being done. In the identification of new genes, development of new drug delivery systems, diagnostics, recombinant vaccines, computational biology, and many other related areas, considerable success has been achieved. Breakthroughs include studies on the three-dimensional structure of a novel amino acid, a long protein of mosquito (University of Poona), and demonstration of the potential of the reconstituted Sendai viral envelops containing only the F protein of the virus, as an efficient and site-specific vehicle for the delivery of reporter genes into hepatocytes (Delhi University).

In India we have achieved the cloning and sequencing of at least six genes, developed regeneration protocols for citrus, coffee, mangrove species, and new types of biofertilizer and biopesticide formulations, including mycorrhizal fertilizers. Research to develop new genetically improved (transgenic) plants for brassicas, mung bean, cotton, and potato is well advanced. Industries have also shown a keen interest in the options of biotechnology and are participating in field trials and pilot level productions. The successful tissue culture pilot plants in the country, one at TERI in New Delhi and the other at NCL in Pune. The world's first IVF buffalo calf (PRATHAM) was born through embryo transfer technology at the National Dairy Research Institute, Karnal. Multiple ovulation and embryo transfer, in vitro embryo production, embryo sexing, vaccines and diagnostic kits for animal health have also been developed.

With a coastline of more than 8,000 kilometers, and two island territories of Andaman and Nicobar and Lakshadweep, there is great potential for marine resource development and aquaculture.

To achieve an annual target production of 10 million metric tons of fish, scientific aquaculture offers great possibilities. In fact, aquaculture products are among the fastest moving commodities in the world. We have to continuously improve seed production, feed, health products, cryopreservation, genetic studies, and related environmental factors.

It is scientifically well established that an environmentally benign way of ensuring food security is through bioengineering of crops. For the billion people in developing countries, one billion do not get enough to eat and live in poverty. Is there any other strategy or alternative? Biotechnology will provide the new tools to breeders to enhance plant capacity. Since we know that 12 per cent of the world land is under agricultural crops, it is projected that the per capita availability may be reduced from 2.06 hectares to 0.15 hectare by 2050.

Our National Plant Genome Research Centre at Jawaharlal Nehru University dealing with plant molecular biology in different parts of the country were initially responsible for training significant numbers in crop biotechnology. There are innumerable possibilities of producing more proteins, vitamins, and pharmaceuticals from plant. Biotechnology is becoming a major tool in conservation biology. Twelve per cent of the vascular plants are threatened with extinction. Over 5,000 animal species are threatened worldwide, including 563 Indian species. India also has about 2000 species of vascular plants that are threatened.

Globally, there are about 35–40 biotechnology-derived therapeutics and vaccines in use and more than 500 drugs and vaccines in different stages of clinical trials. Every year about 12 million people die of infectious diseases. The main killers according to WHO are acute respiratory infection, diarrheal diseases, tuberculosis, malaria, hepatitis, and HIV-AIDS. There are vaccines being developed for many diseases, and diagnostic kits for HIV, pregnancy detection, and hepatitis are being developed. The technologies have been transferred to industry. National Brain Research Centre is being established to improve knowledge of the human brain and the brain diseases. The age-old system of Ayurveda practiced in India needs to be popularized and made an integral part of health care. The global market for herbal products may be around US$5 trillion by 2050. The coming together of biotechnology and informatics is paying rich dividends. Genome projects, drug design, and molecular taxonomy are all becoming increasingly dependent on information technology. Information on nucleotides and protein sequences is accumulating rapidly.

The number of genes characterized from a variety of organisms and the number of evolved protein structures are doubling every two years. In India it has established a national Bioinformatics Network with different Distributed Information Centres which help to establishment of genomic databases. It is also helping to the Indian scientists. There are about several approved MS, postdoctoral, and MD training programs in biotechnology in progress or just about to start, in different institutions and universities covering most Indian States. Short-term training programs, technician training courses, fellowships for students to go abroad, training courses in Indian institutions, popular lecture series, awards, and incentives form an integral part of the human resource development activities in India.

States are taking a keen interest in developing biotechnology-based activities. The States of Delhi,Uttar Pradesh, Arunachal Pradesh, Madhya Pradesh, Kerala, West Bengal, Jammu and Kashmir, Haryana, Mizoram, Punjab, Gujarat, Meghalaya, Sikkim and Bihar have already started large-scale demonstration activities and training programs.

India will require a strong educational and scientific base, clear public understanding of the value of new biotechnologies, and involvement of society in many of these biological ventures.

India has a large research and educational infrastructure comprising 29 agriculture universities, 204 central and state universities, and more than 500 national laboratories and research institutions.

It should therefore be possible to develop capabilities and programs so that these institutions act as regional hubs for the farming community, where they can get direct feedback about new technological interventions. It will be equally important to establish strong partnerships and linkages with industry, from the time a research lead has emerged until the packaging of the technology and commercialization are achieved (4-6).

Advancement in Drug Development, Pharmaceutical Industry and Medical Sector Throughout Globe

The biotech industry has produced several enabling technologies over the years to support faster, cheaper and more targeted drug discovery and development. Ultimately, the goal of these technologies is to reduce (potentially halve) discovery time and increase (potentially triple) the number of new chemical entities (NCEs) entering the development stage. The question is, how will these technologies contribute to the development of better and/or safer therapies for the future? Simply supplying more targets or more drug leads does not, in itself, assure success. What will increase the chances of success– and what some biotech-derived tools will facilitate–is the ability to define, with a high level of certainty, a population that will respond to a particular therapy, with minimal risk for failure. This emerging discipline has been termed 'Pharmacogenomics'(7).

At the heart of Pharmacogenomics are technologies designed to identify single-nucleotide polymorphisms (SNPs), which are small genetic variations that might be linked to disease susceptibility or to drug response. The most prominent use of SNP technology in the future is likely to involve defining populations of potential drug responders compared with potential non-responders for clinical trials, identifying individuals who are likely to experience adverse events in clinical trials or carrying out post-marketing analysis of drug safety and efficacy (8). In fact, the US Food and Drug Administration (FDA) has already voiced support for this approach (9), providing validation for Pharmacogenomics analysis as a valuable tool in the drug-approval process. So far, SNP analysis has confirmed the previous discovery(10-14) of the E4 allele of apolipoprotein (*APOE4*) as the susceptibility gene variant that is linked to common, alzheimer's disease (12), and allowed the discovery *SLC12A84* (16) and the insulin receptor (*INSR*) gene on chromosome 19 (17) as susceptibility gene variants in psoriasis and aurapositive migraine, respectively.

Although not yet adopted by most companies carrying out clinical trials, Pharmacogenomics has yielded valuable information for big pharma companies, such as Glaxo SmithKline (GSK), which has implemented this type of analysis in clinical trials (8). One example of the successful use of Pharmacogenomics analysis–not in the initial clinical trial process, but in post-launch safety analyses–was in a drug-rescue programme for GSK's irritable bowel syndrome (IBS) drug Lotronex (alosetron). The company voluntarily took Lotronex off the US market in 2001 after reports of serious adverse events–that is, ischaemic colitis and other severe complications–some of which resulted in death.

Subsequent Pharmacogenomics studies completed by GSK allowed the company to submit a supplemental new drug application (sNDA) to the FDA for a new and more narrowly defined subset patients in whom the risk/benefit ratio was favourable. The FDA has now approved the drug with the condition that it be prescribed according to a strict risk management programme, and Lotronex was relaunched mid-2002. This decision makes Lotronex the first example of a drug that is being successfully reintroduced to market with a risk-management programme.

A further evolution in this technological approach will be its use in preventive medicine. It is quite conceivable at this time that pharmacogenetics will allow the proactive identification of populations at risk or predisposed towards particular diseases–for example, type 2 diabetes, hyperlipidaemia and high blood pressure. This will give rise to a whole new class of drugs that are designed to stave off the onset of acute disease. The pharmacoeconomic benefit and impact to society would be enormous, as the end-stage co-morbidities that are associated with many of these diseases (18).

Disciplines such as functional genomics and proteomics will serve to expand greatly our understanding of the biological roles of different genes and proteins, both in healthy and disease states.With the map of the human genome completed, functional-genomic studies are rapidly expanding the list of potentially therapeutic genes. Proteomics, which is still a fairly undeveloped discipline in terms of its potential to contribute to drug development, is rapidly maturing, and is widely anticipated to contribute to major advances in drug discovery(19). Understanding the proteome in all its complexity is the new frontier in disease biology, with important clinical ramifications, as it will increasingly refine our molecular understanding of specific diseases (20). As important as genomics is to drug discovery today, proteomics will one day supersede it in its importance in both drug discovery and the practice of clinical medicine.

The Future Revolutions Throughout the Globe

To meet the broader needs of the health-care market in the future, biotech companies are taking novel approaches to the treatment of disease outside the realm of traditional drug discovery and development. Many are pursuing stem-cell biology for use in a wide range of therapeutic applications, whereas others are refining tissue-engineering technologies to aid in the development of whole-organ replacement. Although these approaches will not become a clinical reality in the near future, we predict that they will substantially affect the market within the next 20 years. Stem-cell therapies, in particular, could potentially address a broad range of disease states (21), as the technologies that are being developed in this area could allow differentiation of stem cells into numerous cell types. Recent findings of adult stem cells are enabling tissue and organ engineering (22) that will one day allow wholesale transplantation of new tissues and organs. Discoveries about the growth and differentiation factors and conditions that drive tissue development could, in the future, allow the *in situ* regeneration of cells and tissues within patients.Although this field remains in its infancy, new technologies will

Figure: Evolutionary Tree and Advancement of Biotechnology and Biochemistry

2022: Transplantation of organs and tissues from cloned animals.
Disease predisposition tests and predictive medicine incorporated into disease management.

2020: Number of biotech products approved exceeds 500.
General immunosuppressive approved for transplantation.

2015: Ultra highthroughput sequencing chip arrays routine diagnostic testing.

2013: Stem-cell technology used for *in situ* regeneration.

2010: First oral recombinant protein approved.

2012: First antisense drug for cancer approved.

2010: Pharmacogenetic analysis incorporated into clinical-trial designs and post-market analyses.
First gene therapy approval USA.

2007: First cancer vaccine approved in major markets.

2004: First cancer gene therapy approved in China

2002: Development of the commercial protein chip.

2000: The first draft of the human genome is completed.

1998: Vitravene, the first therapeutic agent developed with antisense technology.

1996: Commercialization of Gene Chip technology.

1995: Popularization of combinatorial chemistry.

1994: The first breast cancer susceptibility gene, *BRCA1*, is discovered.
ReoPro, the first chimeric human–murine monoclonal antibody for therapeutic use.

1990: The first clinical trial for gene therapy.
Launch of the Human Genome Project.

1989: Patent awarded for the first transgenic mouse.

1986: OKT-3, the first murine monoclonal antibody for therapeutic use.
Roferon A, the first recombinant cancer therapy (interferon).
Recombivax HB, the first recombinant vaccine for human use.

1983: Amplification of DNA with the polymerase chain reaction (PCR) technique.

1982: Humulin, the first recombinant therapeutic (insulin).

no doubt emerge to allow the isolation, expansion and differentiation of stem cells from various sources. This could, ultimately, lead to widespread therapeutic applications in areas including cardiovascular disease, neurodegenerative disease, metabolic disease, bone disease and repair, and haematological disorders. A similarly novel approach of nearer-term benefit, and representative of how biotechnology has overcome tremendous hurdles to bring truly novel therapies to late-stage clinical development, is gene therapy (23). Since its inception in the 1970s, the field of gene therapy has been fraught with safety concerns, primarily stemming from gene delivery–which historically has necessitated the use of viral vectors–but also from the need to control gene expression.

Despite these obstacles–and an abundance of negative press throughout the years when novel therapies did not live up to their potential or when safety concerns were a cause for halting clinical trials–a number of companies have stayed the course and survived to see their technologies reach late-stage clinical development. In fact, there are several gene therapies in Phase III trials at present, most of them targeting cancer indications. We predict that at least one of these will be approved by 2006. At present, cancer and cardiovascular disease–both leading causes of death and two of the biggest potential markets in the western world–are the main targets for gene therapies. Gene therapies that have reached Phase III clinical development for cancer include Introgen's head and neck cancer therapy Advexin–for which a BLA could be filed as early as 2004; Onyx Pharmaceutical's ONYX-015 for head and neck cancer; and Vical's Allovectin-7 (HLA-B7 gene therapy) for melanoma.Other active areas of development for gene therapies include infectious diseases (such as HIV/AIDS, hepatitis and human papillomavirus) and central nervous system disorders; in fact, the first gene-therapy trial for Parkinson's disease was approved in October 2002, amidst much excitement within the Parkinson's community, with plans to begin the Phase I trial by 2003. In the fourth quarter of 2002, there were more than 600 completed, ongoing or pending gene-therapy trials worldwide; in the United States alone, the FDA had 196 active gene-therapy Investigational New Drugs (INDs) (106 for *in vivo* gene transfer and 90 for *ex vivo* gene transfer). However, in light of some highly publicized safety concerns from clinical trials in recent years–specifically, the two deaths that led the FDA to halt Vascular Genetics' angiogenesis trial using vascular endothelial growth factor 2 (VEGF2), and the more recent reported case of leukaemia in a child involved in the X-SCID (X-linked severe combined immunodeficiency) gene trial–regulatory agencies in many countries are urging extra caution for gene-therapy trials. That said, many in the scientific community, and several biotech companies with expertise in this arena, hold unwavering support for this approach to treating disease. Despite significant obstacles, the reward potential for a successful gene therapeutic is huge, given the diseases that most companies are targeting (such as cancer and cardiovascular diseases). Similarly, the reward potential for stem-cell therapies or other tissue-replacement technologies, which might encounter obstacles much like those encountered by gene therapies, could be huge if these approaches offer effective treatments for degenerative disorders. In essence, novel biotechnology products, such as those previously mentioned as well as others such as cancer vaccines and antisense therapeutics, hold great potential because they might fill major areas of unmet need that are not currently met by traditional pharmaceuticals.

To achieve the goal of self-reliance in this field, India will require a strong educational and scientific base, clear public understanding of the value of new biotechnologies, and involvement of society in many of these biological ventures.

Arther Kornberg, Nobel Laureate, stated: "*Much has been said about the future impact of biotechnology on industrial development, but this does not yet apply to the less developed countries that lack this infrastructure and industrial strength. In view of the current power of biotechnology and its even brighter future, there is no question that the less developed countries must now position and strengthen their status in biotechnology.*"

REFERENCES

1. Berg, C., Nassr, R. and Pang, K. (2002) The evolution of biotech. **Nature Rev. Drug Discov.** (1) 845–846.
2. Decision Resources, Inc. *Biotech Alliances* (2001): *Feeding and Riding the Tiger* (Decision Resources, Inc., Spectrum, Pharmaceutical Industry Dynamics, Issue 6, 22 May 2002).
3. Decision Resources, Inc. (2002) Outlook for the European Biotechnology Industry: Industry and Government Cooperation Key to Continued Growth (Decision Resources, Inc., Spectrum, Pharmaceutical Industry Dynamics, Issue 9.
4. Sharma, M.(2000) India: Biotechnology Research and Development.51-57
5. French, C.E., S.J. Rosser, G.J. Davies, S. Nicklin, and N.C. Bruce. (1999) Biodegradation of explosives by transgenic plants expressing pentaerythritol tetranitrate reductase. **Nature Biotechnology** 17(5) 491-494.
6. James, C. 1998. Global Review of Transgenic Crops: 1998. ISAAA Brief No. 8. Ithaca, N.Y.: International Service for the Acquisition of Agri-biotech Applications.
7. Lindpaintner, K. (2002) The impact of pharmacogenetics pharmacogenomics on drug discovery. **Nature Rev. Discov**. (1), 463–469.
8. Roses, A. D. (2002) Genome-based pharmacogenetics and pharmaceutical industry. **Nature Rev. Drug Discov.** (2) 541–549.
9. Lesko, L. J. and Woodcock, J. (2002) Pharmacogenomic-guided drug development: regulatory perspective. **Pharmacogenomics J.**(2) 20–24.
10. Roses, A. D. (2002) Apolipoprotein E alleles as risk factors Alzheimer's disease. **Annu. Rev. Med**. (47) 387–400
11. Farrer, L. A. (1997) Effects of age, sex, and ethnicity association between apolipoprotein E genotype and Alzheimer disease. A meta-analysis. APOE and Alzheimer Disease Meta Analysis Consortium. **JAMA** (278) 1349–1356.
12. Tang, M. X. (1998) The *APOE4* allele and the risk of Alzheimer disease among African Americans, whites, Hispanics. **JAMA** (279) 751–755.
13. Kardaun, J. W. (2000) Genotypes and phenotypes for apolipoprotein E and Alzheimer disease in the Honolulu-Asia aging study. **Clin. Chem.** (46) 1548–1554.
14. Graff-Radford, N. R. (2002) Association between apolipoprotein E genotype and Alzheimer disease in African American subjects. **Arch. Neurol**. (59) 594–600.
15. Lai, E., Riley, J., Purvis, I. and Roses, A. D. (1998) A 4-Mb high density single nucleotide polymorphism-based map around human *APOE*. **Genomics** (54) 31–38.

16. Hewett, D. (2002) Identification of a psoriasis susceptibility candidate gene by linkage disequilibrium mapping a localized single nucleotide polymorphism map. **Genomics** (79) 305–314.

17. McCarthy, L. C. (2001) Single-nucleotide polymorphism alleles in the insulin receptor gene are associated with typical migraine. **Genomics** (78) 135–149.

18. Tricia N., Carilee B., Rena Nassr and Kevin P.(2003) The further evolution of biotech. **Nature Reviews** (2): 75-80.

19. Burbaum, J. and Tobal, G. M. (2002) Proteomics in drug discovery. ***Curr. Opin. Chem. Biol.*** (6) 427–433.

20. Petricoin, E. F., Zoon, K. C., Kohn, E. C., Barrett, J. and Liotta, L. A. (2002) Clinical proteomics: translating benchside promise into bedside reality. **Nature Rev. Drug Discov.** (1) 683–695.

21. Lovell-Badge, R. (2001) The future for stem cell research. **Nature** (414) 88–91.

22. Bianco, P. and Gehron Robey, P. (2001) Stem cells in tissue engineering. **Nature** (414) 118–121.

23. Chiranjib Chakraborty and Atanu Bhatcharyya (2004): Pharmacogenomics: Approaches to new drug development. Biotech Books, New Delhi, India, pp 140-143.

Advances in Biochemistry and Biotechnology Volume 1 *Pages 10-52*
Edited by: Dr. Chiranjib Chakraborty
Published by: DAYA PUBLISHING HOUSE

1

Recent Developments in Chondrocyte Cultures and Cartilage Engineering

*R.K. Saxena, Garima Sharma & Prashant Mishra**
*Centre for Biomedical Engineering and *Department of Biochemical Engineering and Biotechnology, Indian Institute of Technology Delhi, Hauz Khas, New Delhi – 110016*

ABSTRACT

Chondrocytes play a major role in physiological and load bearing properties of articular cartilage. Articular cartilage tissue is important for normal and friction free movements of synovial joints. Chondrocytes are embedded in an abundant extracellular hydrated matrix comprising predominantly of type II collagen and proteoglycans. The collagens are organized into an insoluble fibrous network of high tensile strength, which contain the soluble deformable PGs that give the tissue its ability to undergo reversible deformation. The biomechanical properties of articular cartilage rely on the biochemical composition and integrity of its extra cellular matrix. The metabolism of articular cartilage represents a fine balance between synthesis and degradation of extracellular matrix. It is a feature of normal growth development and maintenance of articular cartilage in adult. In degenerative joint disease or osteoarthritis, cartilage shows a distinct loss of proteoglycan, marked collagen network fibrillation, increased tissue hydration and loss of

* Correspondent Author

compressive stiffness of the tissue. These changes are followed by enhanced level of tissue degrading enzymes like matrix metalloproteases-1 (MMP-1), matrix metalloproteases-3, (MMP-3), matrix mettaloproreases-7 (MMP-7), Collagenase, Cathepsin D, hyaluronidase, aggrecanase and various other serine proteases. The progressive irreversible loss of matrix proteoglycans result in diminished tissue function and it likely to exacerbate the disease process by exposing underlying bone to greater mechanical forces. Hence chondrocyte tissue culture systems can be employed as a model to study the cartilage metabolism and to elucidate the complex mechanism regulating the synthesis of proteoglycan. The organoid or explant culture is very useful for studying synthesis, turnover and drug responses with respect to the metabolism of matrix molecules. In monolayer culture chondrocytes are grown at low density with addition of serum. After a short culture period chondrocytes dedifferentiate and are fragmented to fibroblast like cells, change the pattern of their proteoglycan synthesis and overgrown by fibroblast like cells.

Another means of growing chondrocytes is cultivation in suspension or semisolid media using agar, alginate and other gel forming substances. The most recent *in vitro* model of chondrocyte culture employs encapsulating cells in alginate beads. This culture system offers multifold advantages over other models. The chondrocytes remain viable and phenotypically stable for months after encapsulating them in alginate beads. A great advantage of alginate bead culture is that it can be dissolved by chelating substances and allows further investigations of isolated chondrocytes in culture medium. Many studies have suggested that cultivation of chondrocytes in alginate stimulate synthesis of type II collagen and large aggrecans. The alginate system has extensively been used to examine the effect of different catabolic/anabolic agents on synthesis and turn over of extracellular matrix of cartilage. The cartilage tissue engineering approaches to cartilage repair are also focused on developing efficient methods of growing chondrocytes in culture medium. There is growing interest in the development of cartilage repair systems, which involve the implantation of autologous chondrocytes into defect sites for the treatment of localized full, or partial thickness lesions in articular cartilage. The chondrocytes isolated from small tissue biopsy are cultured *in vitro* to increase their number. The chondrocytes are expanded in specially designed biomaterials scaffolds (alginate, collagen fibrin etc), which provides them mechanical stability and supports even cell distribution. The cartilage constructs can be implanted at the site of articular cartilage injuries for maintenance and management of synovial joints. Recent developments in chondrocyte culture and its implications in tissue engineering and rehabilitation have been reviewed.

INTRODUCTION

Chondrocytes are unique type of cellular configurations, responsible for the maintenance of interior milieu of weight bearing synovial joint articular cartilage (AC). Articular cartilage covers the entire condylar mating surfaces of movable joints of the body. It is a highly specialized and uniquely designed connective tissue material that forms the smooth, glistening and friction free lining on the subchondral bones. Chondrocytes are embedded in the extracellular ground matrix of AC and responsible for the providing load bearing and load shearing properties during dynamic joint activities. It provides a wonderful friction free environment for complete maneuvers and flawless locomotion. In addition, articular cartilage is also capable of dissipating the mechanical energy so that the joint structure is protected from any harmful damage by an impact force. Physiologically, articular cartilage is virtually an isolated tissue, devoid of all blood vessels, lymph channels, and nerves. It receives nutrition from the synovial fluid and has two primary functions:

1. To distribute joint loads over a wide area, thus decreasing the stresses sustained by the contacting joint surfaces (shock absorption).
2. To allow relative movement of the opposing joint surfaces with minimal friction and wear (Joint lubrication).

It is well known that degeneration/wearing of AC can lead to several disabling problems and some times complete failure of joints. Maintenance of good health of load resistant cartilage by virtue of well-being of chondrocytes, provides an excellent, dynamic and better quality of life for human populations. In view of this vision currently several research studies are in progress for understanding the properties of articular chondrocytes *in vivo* and *in vitro*, in order to provide a suitable therapeutic method for the prevention of onset and control of progression of degenerative joint disease.

Chondrocyte are metabolically dynamic cells (1) and secrets extracellular matrix largely made up of proteoglycans aggrecans (load shearing material, provides compressive strength), and collagens (load baring material, provides tensile strength and holds proteoglycans aggrecans in mass). Also chondrocytes are responsible for assembly, incorporation, organization and turnover of these macromolecules in cartilage (2,3). Thus chondrocytes form the structural and functional units of articular cartilage. Type II collagen is the major form of collagen in articular cartilage together with smaller amount of other collagen like type I, IV, IX and X (4). The triple helix of type II collagen forms the basic architecture of the tissue. The three dimensional fibrilar network of collagen provides tensile stiffness and strength to the cartilage (5). Large molecular weight aggrecans and other proteoglycans are embedded within this fibrous network and provide compressive stiffness to the tissue (6). These interactions are mediated by electrostatic and covalent cross-linking between charged groups on both the species. In normal physiological conditions, chondrocytes maintain fine balance between synthesis and degradation of these components. However, in certain pathological conditions such as osteoarthritis, this equilibrium gets disturbed in response to various systemic and local etiological factors like aging, trauma, mechanical stresses, obesity and genetics. Pathological cartilage shows a loss of proteoglycan (7), marked collagen network fibrillation (8), increased tissue hydration and loss of compressive stiffness. These changes are followed by altered level of tissue degrading enzymes like matrix metalloproteases-1 (MMP-1), matrix metalloproteases-3 (MMP-3), (9), matrix mettaloproreases-7 (MMP-7), (10) collagenase, cathepsin D, hyaluronidase, aggrecanase (11) and various other serine proteases. These disturbances lead to gradual cartilage degeneration (12). An understanding of the factors mediating cartilage breakdown and their regulation, is of therefore great importance, since from them an approach can be made to design therapies for the reduction of pathological cartilage destruction and enhancement of repair in this tissue. Progress to this goal is dependent upon identifying the degradative events and agents, potentially active in articular cartilage tissue. A number of *in vitro* studies on cartilage have investigated various aspect of regulation of biosynthesis of cartilage matrix, its processing and turnover. Investigators have tried to develop the *in vitro* models of cartilage employing isolated human chondrocytes maintained in culture medium and also with intact articular cartilage tissue culture (13). The aim of these studies is to define the unique metabolic properties of chondrocytes and to develop culture conditions that accurately maintain the cartilage phenotype (14-16). Among the various *in vitro* model systems available for chondrocyte and cartilage culture, includes, cartilage organ culture (17), isolated chondrocytes culture in agarose (18-19), alginate (20-22) and collagen gel (23,24), and as high-density monolayers (25).

In this review we have presented an overview of various chondrocyte culture techniques, and their advantages and disadvantages with respect to the desired application. Also the focus has been given on discussing the current strategies of chondrogenesis and recent advances of *in vitro* cartilage regeneration in direction of providing functional cartilaginous constructs that can be used for controlled *in vitro* studies of chondrogenesis and potentially for *in vivo* cartilage repair, and restoration of joint function in degenerative joint disease like osteoarthritis.

Chondrocytes: The Architect of Articular Cartilage

Chondrocytes are specialized cells, which produce and maintain the extracellular matrix of cartilage, a tissue that is resilient and pliant. Although the number of cells in unit volume of articular cartilage is small in comparison to other tissues (1-10% of the tissue volume) (26), the chondrocytes show considerable metabolic activity (1). The activity is associated with a continuous physiological turnover of the extracellular ground substance in adult articular cartilage, involving its synthesis and degradation.

Articular cartilage is devoid of nerve supply and is generally considered to be an avascular tissue. The cartilage cell thus provide the body with a metabolically economical load bearing material, and the investigators, with interesting problems in understanding life in an avascular environment and modes of synthesis and degradation of matrix molecules such as collagen and proteoglycan. The unique nature of the chondrocytes lies in its ability to form and regulate an organized tissue at sites remote from blood vessels, nerves and lymphatics.

Chondrocytes are one of the few cell types that can survive without being close to a blood supply. However, once the cartilage is damaged, the chondrocytes' distance from a blood supply means that they lack the ability to regenerate themselves, as this process requires an increased amount of nutrients and access through the blood stream and secretion of regenerative proteins. Chondrocyte is surrounded by a specialized matrix distinctly different from the bulk of the intercellular matrix, which makes up more than 90% of the tissue volume. The chondrocytes are surrounded by a specialized matrix to form a structural unit named as *chondron*. Each chondron consist of a 'chondrocyte', its 'lacunar space' and a 'pericellular rim', (27) which represent the primary structural, functional and metabolic unit responsible for cartilage matrix homoeostasis.

The Physical Environment of the Chondrocytes

The chondrocytes of adult cartilage lie isolated from each other and are embedded in a matrix consisting mainly of collagen network cross linked within high concentration of polyanionic proteoglycans. The total ionic composition of extracellular fluid in cartilage is higher due to presence of high concentration of free cations and low concentration of anions relative to external medium like synovial fluid (28). Also the presence of fixed negative charges on proteoglycan ensures that the pH in loaded cartilage is around 0.5 units lower than the synovial fluid pH (29). In addition the chondrocytes obtain their energy by anaerobic respiration therefore the concentration gradient of lactic acid and other metabolites also exist at the articular cartilage joint surface varying from 6% O_2 to less than 1% in the deep layers (30).

In addition to control the ionic concentration and pH of the cartilage extracellular medium, proteoglycan also play their role as selective permeability barrier for entry of large molecules like immunoglobulins and serum albumin and allow very restricted entry of small molecules like cytokines and interleukins in cartilage (31). Thus an intact cartilage chondrocytes are exposed to very low concentration of circulating factors than most of the other cells.

The Chondrocyte: A Cell Under Pressure

Articular cartilage has unique material properties that enable it to perform its physiological functions over a lifetime and under a wide range of loading conditions. Articular cartilage is subjected to high compressive and shears stresses, and during normal activity pressure may rise to 100-200 atm (10-20 Megapascal, MPa) within milisec (32). A biomechanical study performed on cadaveric limbs loaded in stimulated gait have shown that human hip experiences stress in the range of 5-10 Mpa that are 300-800% of normal body weight (33-35). During normal walking the load may range form 2.5 times bodyweight (36) to 5 times bodyweight (37). Thus *in vivo* chondrocytes are constantly exposed to high mechanical loads, which depend on body weight and vary with posture and joint activity. Duration of loading cycle is another very critical factor, as it alters the extracellular environment of the cartilage in a complex manner.

Thus it is clear that resilience and endurance capacity of articular cartilage in physiological condition are determined by the loading pattern and depends critically upon the structure, composition and integrity of its extracellular matrix. The cyclic stress and strain generated during the repetitive load in joint motion may cause fatigue failure within the tissue. The mechanical load not only affect the material properties of the cartilage matrix, but it has been known that chondrocytes also respond to mechanical forces and within limits are able to remodel cartilage in line with the prevailing level of stresses (38). Thus chondrocytes always function in a demanding environment and their metabolic activities get altered in response to mechanical loads. Over the past 40 years, there has been ample demonstration of a surprisingly active level of metabolism in articular cartilage, and it is now clear that the chondrocytes, despite their sparse appearance in adult cartilage, synthesize and assemble the matrix components and direct their distribution within the tissue.

Various Models for the Study of the Articular Cartilage

Since a long time animal models are widely used to investigate the interaction between mechanical and biological factors, to study the metabolic response of cartilage (39-41), and to develop and evaluate tissue-engineering techniques for reconstruction of damaged articular cartilage. The aim of these studies is to engineer articular cartilage for the functional restoration of a joint surface that appears anatomically, histologically, biochemically and mechanically similar to the original joint surface. There are various critical issues in the selection of an appropriate animal model, such as the problems related to availability of the animal, age of the animal, the benefits and limitations of the model systems, scaffold selection with emphasis on polymers, and the most important one is, evaluation of the tissue-engineered articular cartilage in such models. As these models have uncontrollable biological processes so inference of the results is another critical aspect of these studies. Moreover, *in vivo* models have proven valuable only for studying the overall gross tissue response but they are limited in addressing specific questions concerning cellular response to mechanical stimuli and the process of chondrogenesis during the cartilage repair and regeneration. These concerns are now being addressed through the use of new and innovative *in vitro* model systems. Much of the information currently available on cartilage mechanobiology and biochemistry has been realized from tissue culture experiments. There are three basic approaches for this: (1) the intact synovial joint, with all the biomechanical forces placed on it (2) the cells in their native matrix in tissue culture; and (3) culture of chondrocytes after releasing them from the extracellular matrix.

The culture techniques, employing cartilage slices or isolated chondrocytes have proven to be very valuable in studies of metabolic activity and regulation of proteoglycan, secretion, its assembly in aggregates and their turnover (23,24), determining the effect of mechanical, hormonal and autocrine factors on proteoglycan homeostasis. Also, the regeneration of functional articular cartilage by formation of neocartilage is another most tantalizing prospect and intimidating challenge for bioengineers today. The tissue engineering approaches to cartilage repair are focusing on developing efficient methods to form functional cartilage via expansion of chondrocytes obtained from various sources, in synthetic medium and implantation at the defect sites for treatment of localized full or partial thickness lesions in young adults (42-45). As the cartilage is a sparsely cellular tissue so obtaining sufficient number of healthy chondrocytes to grow cartilage *in vitro,* remains one of the greatest challenges in healing injured or degenerated cartilage. The cells may be obtained through culturing primary autologous/herterologous chondrocytes (43), mesenchymal stem cells (46) and embryonic stem cells (47). There are various ways of expansion of chondrocyte population as: monolayer culture, during this process chondrocyte are known to dedifferentiate and adopt a fibroblastic phenotype and lose their characteristic pattern of matrix protein expression, and the other one is chondrocyte culture technique using scaffold material. In the later approach chondrocytes are seeded into a biomaterial, which can be derived from natural (collagen, fibril, hyaluronic acid, alginate) or synthetic sources (poly ethylene glycol, poly lactic acid) (48) Also the scaffold free approach have been used where chondrocytes are cultured at high densities in the absence of any supporting material (49).

There is a hierarchy of culture systems that can be employed as models in studies of cartilage metabolism:

Organ/Explant Culture System

This is the most extensively studied model system in which cartilage explants from calves, steers, pigs and humans have been used for culturing of cartilage and have provided a wealth of information relating to the mechanical and biochemical regulation of cartilage metabolism (50-54). These studies are invaluable because they provide the window through which the behavior of chondrocytes in their native extracellular matrix can be studied.

In organ or explant culture, cartilage together with its extracellular matrix is grown *in vitro* and the chondrocytes function in maintenance mode *i.e.* they do not proliferate, rate of collagen metabolism is very low whereas proteoglycan metabolism is in homeostasis. As expected the articular cartilage explants can be maintained in steady state for several weeks under appropriate nutritional conditions (55).

The aided advantage of explant culture system is that it preserves the immediate pericellular microenvironment of the chondrocytes with its macromolecules including chondrons, cell-binding proteins like fibronectin and laminin, and organization of the collagen fibrilar network in the territorial matrix surrounding the chondrocytes. Thus, it permits a more realistic picture for evaluation of metabolic signals transmitted from and to the chondrocytes, through the complex architecture of the tissue. The organ culture system has been also used to examine incorporation of newly synthesized labelled molecules into cartilage matrix and their turn over (56).

As in cartilage explant culture; chondrocytes are maintained in homeostasis similar to *in vivo* condition, this has inspired various researchers to choose this model system to investigate different aspects of proteoglycan metabolism in further detail. The studies concerning the regulation of proteoglycan synthesis, its turnover (56-58), and response of cartilage to damage by proteinases (59) lipopolysaccharide (58) and interleukin-1 (56, 60, 61) have been extensively done. Also it has been studied that organ culture maintain constant level of collagen type II for several weeks (58,62).

The cartilage organ culture model is ideal for investigating the effects of various agents on cartilage metabolism. It is of particular interest to assess the ability of therapeutic agents to protect cartilage from damage, and to enhance the repair process in damaged cartilage. Morales and Hascall (63) have presented various studies of the effect of fetal calf serum (57), Insulin like growth factors IGF-1 (64), IGF-2 (65), TGF-β (66), IL-1 (67) and hyaluronic acid (68), on proteoglycan homeostasis in cartilage explants of bovines. Cartilage explant studies have also been performed to investigate the interaction between mechanical and biological factors including the effect of unloading, static loading and overloading (41, 69-71). The major challenge in organ culture is to assure sufficient gas and nutrient exchange in order to maintain viability of cells throughout the tissue mass.

A disadvantage of the organ culture system is that extraction of newly synthesized aggregating proteoglycan from cartilage is a difficult process and requires the use of nondenaturing solvents (72). The presence of matrix containing large anionic molecules makes the process of extraction and analysis of m RNA and protein, exceedingly difficult. In addition the explant system is not suitable for studies related to the cell surface receptors etc.

Isolated Chondron Culture System

Another alternative for studying chondrocyte metabolism is to isolate and culture chondrons. One or more chondrocytes in articular cartilage are surrounded by a pericellular matrix (PCM), which consist of type VI collagen and proteoglycan and together with the chondrocyte have been termed as the 'chondron' (27,73). The pericellular matrix has been hypothesized to play important roles in regulating the biomechanical, biophysical, and biochemical signals that the chondrocyte perceives during normal joint activity. In a recent study, Graff (74) reported that the presence of native pericellular matrix in chondron enhanced the development of mechanically functional neocartilage in porcine pellet cultures. Histological and contemporary arguments suggest that the chondrons act as a fluid filled compression resistant bladder, which seems to protect chondrocytes function by dampening the mechanical, physicochemical and osmotic changes, induced in the matrix by dynamic loading of joint.

Chondron culture has received very limited attention because of the limited yield and viability after mechanical isolation of chondrons from cartilage tissue. Studies to date have focused on developing a variety of techniques to exploit the isolated chondrons as a model of cell matrix interactions in articular cartilage. The isolated chondrons model offers the potential to study dynamic chondron function independently of the bulk extracellular matrix, and development of techniques in this area is vital for understanding articular cartilage function.

Monolayer or Two Dimensional Chondrocyte Culture Systems

The cartilage engineering and autologous chondrocyte implantation approaches are based on expansion and culturing of cartilage cells removed from their native extracellular matrix. The treatment of traumatic or congenital cartilage defects is limited by the fact that cartilage has no intrinsic repair mechanism and the amount of cartilage available for autologous transplantation is always limited. Currently, chondrocyte expansion using monolayer culture is most widely used technique in cartilage tissue engineering. For isolation of chondrocytes from cartilage, the tissue is enzymatically digested with various matrix-degrading enzymes like hyaluronidase, pronase and collagenase. The chondrocytes, thus released from the matrix are maintained in cell culture medium, where they proliferate and lay down new matrix. Studies have confirmed that under appropriate culture conditions the chondrocytes display their *in vivo* phenotypic properties like spherical shape, synthesis of type II collagen and cartilage specific proteoglycan (75,76).

There are two methods of cultivation of chondrocytes isolated from articular cartilage:

Low-density Monolayer Culture

In low-density monolayer culture, after separating the chondrocytes from cartilage, they are grown at low density (3500-5000cells/cm^2) in presence of serum. However various studies have suggested that *in vitro* multiplication of chondrocytes in low-density monolayers result in dedifferentiation of these cells. They lose their specific phenotype and become more fibroblast-like (77-79). Production of the specific cartilage matrix components collagen type II and aggrecans is gradually shifted toward collagen type I (80-82), and small proteoglycans and also alteration occurs in pattern of chondrocytes surface receptors proteins (83). Finally the chondrocyte are overgrown by fibroblast-like cells (84).

High-density/Organoid Culture

The chondrocytes are seeded in high density (<30,000 cells/cm^2) in presence of fetal bovine serum. Various studies have shown that in high-density culture chondrocytes remain stable for 2-3-weeks period and then either transformed to fibroblast like cells (85-87) or to hypertrophic cartilage cells with mineralization of matrix (88,89). One disadvantage of high-density culture is that it does not yield pure cartilage specific cells but a mixed population of cells with perichondrium and nonspecific connective tissue material between the cartilage nodules (83, 86, 87, 89).

In a recent study (90), attempts have been made to regulate the cultivation of chondrocytes from mouse embryo cartilage in monolayer culture and slow down their transformation or their overgrowth by fibroblast-like cells, employing b1 subfamily of integrin receptors, which mediate chondrocyte-matrix interactions. This interaction can be regulated or synergized by growing chondrocytes on collagen type II coated surface in presence of growth factors, such as IGF-I, which stimulates many chondrocyte functions. The results of this study are very promising as cells maintained their round phenotype till the end of cultivation and a higher cell density in comparison to control. The cartilage-specific surface receptors (integrins of the b1-group) could also be demonstrated on the membrane of these cells. Hence, collagen type II and IGF-I prevent chondrocyte dedifferentiation to fibroblast like cells and this model allows a pure chondrocyte culture.

Monolayer chondrocytes culture has been useful in several studies concerning the effect of soluble factors like cytokines, on cartilage metabolism. The technical procedures of culturing and analysis are simple and many variables are relatively well controlled. The disadvantage of phenotypically different population of cells is sometimes less considered. Monolayer, or two-dimensional, culture of disaggregated, anchorage-dependent cells *in vitro,* has allowed progress in understanding regulation of cell differentiation, growth, and function as well as cell-to-cell interactions that modulate these processes.

Monolayer chondrocytes culture with high cell density are useful for studying the intracellular events involved in proteoglycan biosynthesis (91) and in some cases, the process of proteoglycan aggregation (92). Only very little cell associated matrix get accumulated in these cultures, and the chondrocytes secret newly synthesized macromolecules more or less directly into the medium. As a consequence, proteoglycan and other matrix components synthesized in the presence of radiolabelled precursors are easily recovered from the medium in relatively large amounts with a high specific radioactivity (93). For the same reason monolayer cultures are not useful for examining proteoglycan turnover or for studies of mechanism and molecules involved in matrix assembly.

Suspension or Three Dimensional Culture System

A most popular means of growing chondrocytes is cultivation in semisolid media. Chondrocytes grown in solid materials or gels like agarose, alginate and collagen, maintain their spherical shape, reestablish an abundant cell associated matrix and permit the maintenance of their differentiated state. These culture systems are most relevant models as they reflect how chondrocytes are seeded onto various scaffolds; the approach of using scaffolds is a very important aspect of cartilage engineering.

A variety of matrixes have been used to culture chondrocytes three dimensionally, mainly employing natural materials like agarose, alginate, collagen and fibrin; and also synthetic materials like Poly glycolic acid (PGA), Poly lactic acid (PLA) and polyurethanes etc. In particular, agarose and alginate represent the two most widely employed gel systems, and have been extensively used to examine the effect of different agents on synthesis and turn over of proteoglycan by articular chondrocytes of several species (14,16,19,21,72,94). These *in vitro* models are also useful in comparing the properties of specific cell populations from different regions of articular cartilage from different joints (95). In addition, gel suspension culture can be used as a source of matrix-embedded chondrocytes to be tested in biomechanical experiments. Gel containing suspended chondrocytes with their newly formed and laid down matrix can be loaded by varying loads in order to determine the stress-strain relationship (19) of chondrocytes and provide a useful tool in studying the etiology of osteoarthritic like disorders.

Agarose-Chondrocytes Suspension Culture

The entrapment of chondrocytes within agarose (14,18,96) or on agarose (97) has become a popular culture system. Most of the proteoglycans, which are synthesized by the cells and are retained in the vicinity and accumulate with time, forms a matrix that appears to share many of the physicochemical properties of normal cartilage matrix (19). Studies have confirmed that chondrocyte culture in agarose gel maintain the expression of aggrecans (cartilage specific large aggregating proteoglycan) and collagen types characteristic of articular cartilage (predominantly type II, also type IX and XI) (18,72,98). The maintenance of chondrocyte phenotype during long-term culture in agarose has shown the development of a mechanically functional cartilage like matrix (19) Also chondrocytes in agarose gel respond biosynthetically to static and dynamic mechanical load in a manner similar to that of intact organ culture (99). A significant advantage of this culture system is that the cartilage matrix as collagen type II and keratan sulfate, produced by chondrocytes, can be separately characterized using techniques of immunohistochemistry (100).

Alginate-Chondrocyte Suspension Culture

Alginate bead system was firstly used by Jourdian and colleagues (20), to study the metabolism of proteoglycan in human articular chondrocytes. Since then alginate gels have been extensively studied for cartilage tissue engineering applications as a matrix for cell encapsulation and culture. Suspension of chondrocytes in alginate gel promotes the retention of the chondrocytes phenotype and studies have demonstrated that bovine articular chondrocytes suspended in alginate beads remain phenotypically stable for as long as 8 months (16,21), and synthesize cartilage matrix components, including type II collagen and aggrecan (101).

Alginates are naturally derived polysaccharides block copolymers, composed of regions of sequential β-mannuronic acid (M-blocks), and α-L-glucoronic acid (G-blocks) monomers. The length of the M- and G-blocks and their sequential distribution along the polymer chain, varies depending on the source of the alginate (102). Alginates undergo reversible gelation in aqueous solution under mild conditions through interaction with divalent cations such as Ca^{2+} that can cooperatively bind between the G-blocks of adjacent alginate chains creating ionic interchain bridges (103).

Interestingly, the alginate bead culture system has unique property that makes it possible to study the accumulation and turnover of macromolecules in two distinct compartments of cartilage matrix. The matrix that rapidly forms around the cells has been termed as cell-associated matrix (CM), and believed to correspond to the pericellular and territorial matrix compartments, and further removed matrix (FRM), which corresponds to the interterritorial areas of the cartilage matrix, forms further away from the cells (13,104). Using this unique property of alginate culture systems, recently a study has been done on chondrocytes of different developmental stages, to visualize the age related differences in the accumulation and rate of synthesis of proteoglycan, hyaluronic acid and collagen in the two matrix compartments (105). These studies performed on chondrocytes isolated from articular cartilage of young and adult bovine animals, suggested that alginate system could be an effective model in determining the biochemistry of age related cartilage changes like osteoarthritis.

Another very important feature of alginate system is that the beads can be quickly solubilized by chelating substances like ethylene diaminetetraccetic acid (EDTA), and phosphates, and cells can be released after a certain time in culture. The aggrecans and collagen molecules can be readily extracted from each compartment and analyzed further. Thus alginate culture system facilitates the use of experimental approaches not accessible while culturing chondrocytes in other systems like monolayers, explants or agarose gel.

Although alginate culture keeps chondrocytes in a spherical confirmation, supports differentiated phenotype and synthesis of cartilage specific markers, yet chondrocytes suspended in alginate do not adhere to the matrix as they would *in vivo*. For this an important modification in the protocol of using alginate matrix for chondrocyte culture has been to encourage the attachment of cells to the alginate beads via integrins and other cell surface receptors. A most recent study, which has been carried out by Genes (106) deals with designing a modified alginate surface by covalent attachment of GGGG**RGD**Y peptide sequence to low viscosity alginate. The hypothesis is that RGD modification of alginate enables chondrocytes to attach to them via integrins, which are mechanoreceptors and can sense changes in the mechanical environment of the matrix to which they are attached.

The histological and electron microscopical studies of osteoarthritic chondrocytes in alginate culture indicated that after three passages in monolayer, chondrocytes display typical round morphology (107). Study has also demonstrated that human osteoarthritic chondrocytes embedded in alginate beads after their culture in monolayer for three passages developed a pericellular matrix enriched in keratan sulfate and chondrotin sulfate glycosaminoglycans. Consequently, the alginate system appears to be an ideal model for the study of cartilage matrix components in a three dimensional system.

Despite these interesting findings the morphometric and biochemical characterization of the different matrix compartments of chondrocytes cultured in alginate needs to be established further; and it has yet to be shown that how these two compartments can be readily separated at various time points in culture.

Collagen-Chondrocyte Sponge Culture

A relatively recent approach in three-dimensional culturing of chondrocytes, is to seed them with porous collagen sponges. Histochemical and immunohistochemical analyses have confirmed the presence of cartilage specific matrix in cells (108). Analysis of the cartilage-specific genes, aggrecan and collagen type II; have demonstrated the preservation of the chondrocyte molecular signature compared to dedifferentiated monolayer culture (109). Preliminary investigations indicated that articular and endochondral chondrocytes retain their differences in these collagen sponges (110). Also the collagen sponge has been used to determine the effects of insoluble, soluble, and mechanical factors on chondrogenesis. In this direction attempts have been made to study the effect of hyaluronan by incorporating it in collagen sponge. It was found that small amount of hyaluronan (2% w/w) in 3D collagen scaffolds enhanced chondrogenesis, but a greater amount was found to be inhibitory (111).

This finding lends support to the idea that biomimetic, or "smart" scaffolds can be developed with optimal signals for specific histogenesis.

Micro Carrier Suspension Culture

Attempts to maintain the original phenotype of chondrocytes is the most challenging aspect of cartilage engineering and its regeneration. For this attempts have been made in monolayer culture by seeding chondrocytes at high density and to entrap cells in a three-dimensional matrix or scaffold such as agarose and alginate beads, as already been discussed. Despite these modifications in culture conditions, not all chondrocyte population reverted to the spherical, type II collagen synthesizing cells and therefore these methods are not able to provide sufficient numbers of chondrocytes with original phenotype needed for engineering the neocartilage. There is need for a culture system that would allow chondrocytes to proliferate as well as maintain their original characteristics. For this the most recent technique of culturing cells that has been shown to facilitate cell proliferation as well as maintain synthesis of cellular products is the 'microcarrier suspension culture'. This culture system is based on the original 'bioreactor' design to grow bacteria in large quantities. The same principle is used to expand anchorage-dependent mammalian cells in microcarriers and thus mimic other cells types that grow in suspension culture (112). Under controlled levels of pH, oxygen, nutrient supply and mechanical agitation more than 80 anchorage-dependent cell types have been propagated using various types of microcarriers like glass beads, charged molecules such as dextran, sepharose and cross-linked collagens (113-116). Frondoza (117) studied the proliferation and production of cartilage matrix components in human articular chondrocytes in microcarrier suspension culture. Various types of microcarriers, which were used in the study are: cytodex 1 (180pm dextran beads); cytodex 2 (155 pm dextran beads); cytodex 3 (175 pm denatured pig skin type I collagen-coated dextran beads); and cellagen (collagen type I derived beads). From the study it was concluded that out of the four microcarriers, cellagen beads, which are made out of collagen type I, were the most suitable, promotes cell attachment, cell proliferation and allows easy retrieval of chondrocytes with mild trypsin treatment. The retrieved chondrocytes were metabolically active and can be utilized for further experiments.

Recent studies indicated that microcarrier suspension culture system enhanced the growth of bone cells and chondrocytes from rat calvaria proliferating in cytodex 1 microcarriers and continued producing alkaline phosphatase and other cellular products (118,119). These studies have confirmed earlier observation that microcarrier suspension culture has the advantage of increasing division rates of chondrocytes. Thus, limited number of chondrocytes obtained from cartilage could be expanded to greater numbers over a shorter period of time. Studies have also documented success of repairing cartilage defects with chondrocytes grown in microcarrier suspension culture (43). Mass cultures of chondrocytes retrieved from microcarriers may be made available for cartilage regeneration.

It is apparent from the above discussion that interest in the application of gel suspension culture system for *in vitro* proliferation of cartilage has grown tremendously. The potential advantages of these cultures are not restricted to the fact that the cell environment is physically stabilized and that the cells kept well in place after transplantation. These three dimensional systems, as outlined in the sections above, offer possibilities to obtain controlled cell multiplication in a phenotypically stable fashion. It is noteworthy that growth factors like interleukins, cytokines have been shown to stimulate the proliferation of chondrocytes, induce the differentiation of mesenchymal cells into chondrocytes, and stimulate the synthesis of cartilage specific molecules. Improved protocols for the culture of human chondrocytes in gel suspension systems are likely to offer novel opportunities in the treatment of articular cartilage lesions.

Table 1.1: Comparative Study of Different Chondrocyte Culture Systems

Characterstics	*Monolayer*	*Explant*	*Suspension*	*References*
Chondrocyte phenotype	Poor (at low density type I collagen)	Excellent (90% type II collagen)	Very good (in agarose/alginate, 70% type II collagen)	16, 21, 62, 81, 82, 98
Cellularity	High	Low	Medium	16, 21,55, 85,
Expansion Capacity (to increase cell no)	Able to expand (10 X)	No expansion (1X)	Able to expand (10X- 25X)	16, 21,55
Difference in cell population	Has not been shown	Yes	Yes at least to some extent	13,72, 87, 89, 95, 99
Matrix dependent control of cell behavior	No	Yes	Disrupted initially reestablished later	13, 16,51, 86, 91
Penetration of soluble factors	Very good	Poor (uneven)	Good	63, 64, 111
Metabolic activity with respect to matrix constituents	Strongly stimulated anabolism	Steady state with slow turn over	Enhanced metabolic activity, turnover rate similar to *in vivo*	55, 104, 105
Technical handling	Easy	Difficult	Less difficult	93
Quantification of gene products, isolation of macromolecules	Easy to isolate m-RNA	Difficult to isolate m-RNA	Easy to isolate m-RNA (in alginate)	72, 88, 93, 109
Matrix supramolecular organization study	Limited possibilities, not very useful	Possible, but difficult	Possible	56, 57, 104
Nutrition to cells	Good and constant	Poor	Good and constant in many cases	–

Chondrocyte Culture and Cartilage Engineering: Role in *in vitro* Regeneration of Articular Cartilage

In 1743, Hunter stated, *"From Hippocrates to the present age, it is universally allowed that ulcerated cartilage is a troublesome thing and that once destroyed in not repaired"* (120). Several factors may explain why cartilage demonstrates very little capability for self-repair after various injuries. The most obvious one is that chondrocytes are not required to proliferate as in some other tissues (*i.e.* skin), and mature adult chondrocytes have a relatively low metabolic activity (121). Also the articular cartilage is avascular, alymphatic and aneural tissue; and inherent healing of cartilage defect can be achieved only when the defect penetrates the subchondral bone and triggers response from the mesenchymal cells present in bone marrow. Except in the case of full-thickness defects, there is no direct access to progenitor cells and also the resident chondrocytes may be obstructed by the extracellular matrix from migrating to fill the defect site (122,123).

A substantial amount of research has been conducted on articular cartilage, with significant advances in our understanding of the biological repair process being made over the past four decades. While many cartilage repair techniques have been proposed, none have successfully regenerated long-lasting and functional hyaline cartilage to replace the damaged tissue (124). Most of the surgical interventions to repair damaged cartilage have been aimed towards the treatment of clinical symptoms rather than the regeneration of hyaline cartilage, such as pain relief and functional restoration of joint

(125). These approaches include tissue adhesives (126,127), enzymatic treatments (128), laser solder welding (129), autograft cell/tissue transfer via periosteal grafts (130), autologous osteochondral grafting such as mosaicplasty (131) and the carticel method (43). Now the rapidly emerging field of tissue engineering holds great promise for the regeneration of functional cartilage tissue substitutes *in vitro,* for subsequent implantation *in vivo*. Tissue engineering has a primary goal of restoring, maintaining or improving tissue function by employing the principles of engineering and life sciences (132). Tissue engineering of articular cartilage involves the isolation of articular chondrocytes or their precursor cells that may be expanded *in vitro* and then seeded into a biocompatible matrix, or scaffold, for cultivation and subsequent implantation into the joint. The concept of tissue engineering approach in cartilage repair was first proposed by Green (133), clinical applications of this tissue engineering approach was first attempted by a Swedish group (43) with fair-to-excellent results. The success of a neo-cartilage depends upon various factors, which include cell type, scaffold characteristics and culture conditions of isolated chondrocytes in these scaffolds. Stabilization of the differentiated phenotype of the cells and the optimization of neomatrix deposition, composition and organization are crucial factors that have to be considered in cartilage engineering.

The strategy for tissue engineering of cartilage can be classified into six phases:

1. Seeding of bioresorbable polymer scaffold, characterization and toxicity study
2. Seeding of the chondrocytes populations into the polymeric scaffold in static culture (Petri plate)
3. Growth of premature tissue in a dynamic environment (spinner flask)
4. Growth of mature tissue in a physiological environment (bioreactor)
5. Surgical implantation and evaluation *in vivo*
6. Tissue engineered transplant remodeling

Some of the most important aspects of cartilage engineering are described as following:

Identification of Cell Type

For any tissue engineering based method, the critical factors while selecting the cell type are proliferative capacity of cells, phenotypic stability and immunogenicity. However in fully differentiated autologous cells there is no concern of immune response but the issues related to proliferative capacity, donor site morbidity and limited life span, have brought the usage of other cell populations like mesenchymal stem cells and progenitor cells into the forefront of such applications. Various cell types which are being used for expansion and production of neocartilage are described as following:

Differentiated Chondrocytes

Currently the autologous chondrocytes transplantation is the only cell-based method for articular cartilage repair approved by FDA. The procedure involves harvesting of cartilage from healthy regions of joints, their isolation and expansion and subsequent transplantation at the affected site. However the mature chondrocytes have only limited proliferative potential and the number of cell divisions they undergo *in vitro,* decreases with age (134). Differentiated chondrocytes grown as a monolayer loose their native phenotype and dedifferentiate with passaging. However the chondrocyte phenotype may be rescued by use of scaffolds (18,22), reducing oxygen tension (135) and use of growth factors such as basic fibroblast growth factor (FGF-2), transforming growth factor β1 (TGFβ-1) (136) and high concentrations of fetal calf serum (43). The method of autologous chondrocyte implantation is effective in patients with healthy cartilage at other than the focal site of injury, but in the patients with symptoms of arthritis and other joint diseases, own chondrocytes may not be useful and hence an alternative source is required.

Differentiated chondrocytes can also be obtained from non-articular, hyaline cartilage tissues such as nasal or rib cartilage. However, additional work will be necessary to assess whether cells of non-articular origin can be successfully used for articular cartilage repair.

Mesenchymal Stem/Progenitor Cells (MSCs/MPCs)

A stem cell is a non-committed cell from young and adults, and can give rise to specialized cells, tissues and organs. Adult bone marrow is a major source of undifferentiated stem cells (bone marrow stromal cells BMSCs), which contribute to the regeneration of mescenchymal tissues such as bone, cartilage, ligament, tendons and stroma. (46). The multipotent nature of MPCs makes them theoretically ideal candidates for cartilage repair (137). Gao and coworkers (138) tested this hypothesis by attempting repair of osteochondral defects using MPCs loaded in injectable calcium phosphate and hyaluronate derivative. After that Noth (139) also tried *in vitro* engineering of cartilage by using MPCs on PLA scaffolds. This approach provides the added benefits of minimizing handling and maximizing biocompatibility. However, the mechanism of stem cell chondrogenesis is a complicated one and regulated by highly complicated consortium of growth factors, signaling molecules, cell-cell and cell–matrix adhesion molecules. Thus designing the microenvironment of the progenitor cell to obtain the desired functional stage is the most challenging part.

The cells with chondro-osteoprogenitor properties have been isolated from several tissues, including periosteum, bone marrow, spleen, thymus, skeletal muscle, adipose tissue, skin and retina (140-144) and recently in low efficiency, from peripheral blood (145). The use of progenitor cells has not always been straight forward in cartilage engineering for various reasons like hardly accessible tissue source, low frequency, limited information on the particular cell system and lack of understanding about growth requirements of the cells.

Embryonic Stem Cells (ES)

Embryonic stem cells are isolated from the preimplantation embryos, human blastocysts and fetal tissues. Introduction of ES cells in 1998 gave the scientific community an exciting new prospect of cell-based therapies in tissue engineering. Human ES cells can be maintained in culture for more than two years, thus have ceaseless proliferation and pluoripotent nature. However, human ES cells have not yet been reported to be used specifically for musculoskeleton repair due to issues related to ethics, therefore researchers have directed the use of murine ES cells toward the chondrogenic (146) and osteogenic lineages (147) with some success. The culturing of ES cells requires use of feeder layers and technically a cumbersome process. A lot of research is required in this direction to fully understand and optimize the potential of these cells in cartilage tissue engineering.

Coculturing of Chondrocytes/Mesenchymal Cells

Engineering cartilage using multiple cell types is most recent approach and provides the opportunity for cells to provide regulated and synergistic effect on each other for functional tissue development. Culturing of chondrocytes with MSCs induces the chondrocyte expansion by providing paracrine effect of growth factors for chondrogenesis and differentiation of passage cultured MSCs in chondrocytes. The approach also deals with the problem of scarcity of cell source. In a very recent study carried out by Tsuchiya *et al.* (148), human MSCs and bovine mature articular chondrocytes were mixed and cultured under appropriate conditions in pellet culture. The culture showed upregulation of cell proliferation, cartilaginous matrix production and type II collagen gene expression.

The coculture strategy appears to be useful for the *in vitro* expansion of mature chondrocytes without decreasing their differentiation ability. Further trials are in progress to clarify the effect using primary MSCs and autologous chondrocytes cells.

Identification of Appropriate Scaffold

A scaffold provides the cell with a three dimensional matrix upon which it adhere, proliferate and produce matrix components and may be tailored to serve as a vehicle for cell delivery at the site of injury. Scaffold also performs tissue inductive effect by controlling release of growth factors and thus forms functional tissue structure and fills the lesion.

Scaffolds designed for use in cell-based therapies to repair damaged articular cartilage should ideally provide the following characteristics:

1. A three-dimensional (3D) and highly porous structure to support cell attachment, proliferation and extra-cellular matrix (ECM) production;
2. An interconnected/permeable pore network to promote nutrient and waste exchange;
3. A biocompatible and bioresorbable substrate with controllable degradation rates by hydrolytic or enzymatic mechanisms;
4. Minimization of leaching of toxic products like monomers, crosslinkers and solvents;
5. A suitable surface chemistry for cell attachment, proliferation, and differentiation;
6. Mechanical properties to support, or match, those of the tissues at the site of implantation;
7. An architecture which promotes formation of the native anisotropic tissue structure; and
8. A reproducible architecture of clinically relevant size and shape

Selection of scaffold is a considerable challenge in musculoskeleton tissue repair keeping in view the highly demanding environment of load bearing joint. Articular cartilage imposes some special requirement on scaffold material as evident from its unique functional environment, biochemical composition, lack of vascularity and concomitant inability to injury and disease. A variety of biomaterials, autologous and heterologous, naturally occurring and synthetic, biodegradable and non-degradable and bioceramics, have been introduced as potential cell-carrier substances for cartilage repair.

Among the naturally occurring biomaterials are various forms of types I and II collagen-based material, as scaffold matrices (149, 150) gels (151,152) or collagen–alginate composite gels (152), alginate, fibrin, chitosan and glycosmainoglycan based scaffolds. The synthetic degradable biomaterials include polyglycolic acid (PGA) (153,154) and poly-L-lactic acid PLLA (155,156) and PGA–PLLA copolymers (157, 158) etc. Several investigators have also found that some non-biodegradable polymer substances, such as polytetrafluoroethylene (PTFE), (159) polyethylmethacrylate (178), and hydroxyapatite/Dacron composites, (160) also facilitate the restoration of an articular surface in musculoskeleton disorders.

Natural and synthetic biomaterials have their respective advantages and disadvantages. Natural polymers have good cytocompatibility, biaoactivity, cell interaction and hydrophilicity, but they are mechanically too weak to maintain the desired shape and to withstand physiological loading. Synthetic polymers have relatively good mechanical strength and enable precise control over the rate of degradation, porosity and microstructure but their surfaces are hydrophobic and lack cell-recognition signals. Although results of preliminary clinical studies have indicated towards the potential of naturally derived polymers, but concerns about the feasibility of source of these materials for clinical applications, as well as the fear of infection at the site have prompted researchers to investigate the use of synthetic polymers. The synthetic biomaterials can be easily mass-produced and their properties can be tailored for specific applications. There is also the possibility that growth factors or other pharmacologics could be included in the scaffold to encourage certain cellular responses, such as proliferation or differentiation. However only few of these polymers have been FDA approved for use in humans (162).

Some of the most commonly used natural and synthetic polymers, their applications and evaluation in cartilage engineering has been discussed as following:

Natural Scaffold Materials

Collagen

Collagen is the naturally occurring and most abundant protein of the tissue with very high mechanical and shear resistance properties. These properties obviously make it a suitable candidate for delivering chondrocytes at the site of injury. It can be prepared in various forms like solution or shaped into membrane films, threads, sponges and, more recently, into acidic hydrogels. Collagen scaffolds have been already used in *in vitro* characterization of chondrocytes and stem cell behavior study (163, 164). Wakitani *et al.* (165) reported that MSCs in collagen gel brought about the repair of full thickness cartilage defects in rabbit. Collagen matrices containing glycosaminoglycans are also under investigations for inducing hyaline cartilage repair (166).

A most recent study conducted by Ohno (167) has shown that both type I and type II collagen sponge had equivalent effect in cell-seeding efficiency and chondro specific gene expressions. Another study evident that redifferentiation of dedifferentiated bovine chondrocytes occurs when cultured in vitro in a type I collagen- PLGA hybrid mesh (168). These results indicate that the type I collagen sponge is as useful as type II collagen sponge for constructing hyaline cartilage like tissue. Van Susante 1995 (169) demonstrated that type I collagen gels enhance the proliferative capacity of chondrocytes better than alginate beads. Human chondrocytes from non-fibrillated areas of osteoarthritic articular cartilage cultured in a collagen type I microcarrier suspension system, multiplied at least 20 fold within 2 weeks (170). It has also been shown that chondrocytes cultured at high density in collagen type I gels can maintain their phenotype by synthesizing type II collagen and cartilage specific proteoglycans (171).

Collagen matrices are usually derived from xenogeneic sources (mostly porcine and cow skin), the immunological concern can be eliminated by purification of immunogenic telopeptides. Still the possibility of xenozoonoses, the microbial transmission from the animal tissue to the human recipient, has to be carefully considered and remains one of the primary obstacles in cartilage transplantation using collagen (172).

Fibrin

It is a natural polymer that forms during blood coagulation and plays an important role in homeostasis and wound healing. Fibrin glue has been considered for use in a variety of orthopaedic procedures, including fixation of chondral fragments, chondral chips, and osteochondral fragments (173). Fibrin preparations have also been considered as a material for delivering chondrocytes or growing cartilaginous tissue for cartilage repair (174, 175). Chondrocytes interact with fibrin glue in the autologous chondrocyte transplantation procedure, when injected chondrocytes are sealed within cartilage defects by application of fibrin glue to overlying periosteum (176). However, chondrocyte growth and matrix production vary between fibrin preparations (177). In an equine study the incorporation of IGF within polymerized fibrin improved the histology, amount of type II collagen in chondro lesions; however the mechanical properties of the tissue and biochemical features were not consistent with normal articular cartilage (178). Recently Gelse *et al.* (179) implanted MSCs transfected with adenoviral vector containing bone morphogenetic protein–2 (BMP 2) and IGF-1 c-DNA suspended in fibrin glue in rabbit full thickness cartilage lesions and demonstrated that cells were able to attach to the wounded cartilage and cell growth is consistent. Fibrin gel is useful as a model because it is an FDA approved material, and can also be obtained from a patients' own blood (180).

Alginate and Agarose

These two polysaccharides are derived from seaweed and form hydrogels that have the advantage of allowing a uniform distribution of cells throughout implantable scaffolds. The hydrogels have potential to be injected at the defective site and *in situ* form a gel structure with cells encapsulated in them. Agarose has been used extensively in *in vitro* studies of chondrocyte behavior however it does not resorb well, and may elicit a foreign body giant cell response *in vivo* (181). Alginate has been widely used for delivering chondrocytes. Diduch *et al.* (182) reported that alginate beads containing bone marrow stem cells, remained within rabbit osteochondral defects and progressively filled the defects with regenerated tissue. Histological analysis confirmed the chondrocytic phenotype and good amount of proteoglycan synthesis, also binding between the regenerate and host tissue was excellent. A more recent study by Dausse *et al.* (183) however, showed that despite good histological results with cell-seeded alginate, biochemical and mechanical properties of the regenerated tissue were significantly inferior to those evaluated in native cartilage. An alteration of surface properties of scaffold using cell adhesion proteins as RGD has introduced further scope of research of alginate as a carrier in cartilage engineering (106).

Hyaluronic Acid

Hyaluranon is a major component of cartilage matrix. It has unique physico-chemical properties and distinctive biological functions. It binds specifically to proteins in the extracellular matrix and on the cell surface. Partial esterification of hyaluronic acid increases its stability and makes the molecule particularly suitable for cartilage engineering. (184). Butnariu-Ephrat, (185) implanted MScs in a hyaluranon-based glue in goat cartilage defects with little success. Other studies also indicated at much less potential of hyaluranon for cartilage engineering but in a recent study Girotto, (186) evaluated the re-differentiation capacity of cultured chondrocytes on two bioresorbable non-woven biomaterials composed of hyaluronic acid benzyl esters: HYAFF-11, and its sulphated derivative, HYAFF-11S. These experiments have shown that HYAFF-11 and sulphated HYAFF-11 offer an excellent support for *in vitro* chondrocyte culture, generating a functional cartilage construct that expresses specific genes for collagen types II, IX, × and aggrecans. Sulphated HYAFF-11 specially promoted the rate of cell proliferation in both human and chicken cells. Chondrocytes grown in three-dimensional HYAFF-11 scaffolds demonstrated gene expression characteristic of differentiation/re-differentiation. This may prove to be an effective delivery system for chondrocyte cells for the treatment of articular cartilage defects, and an *in vitro* system for studying cartilage differentiation. Further research is on to study human mesenchymal stem cell differentiation towards cartilage generation using sulfated HYAFF-11 scaffolds.

Gelatin-Chondrotin-Hyaluronan Tri-copolymer

Recently a tri-copolymer scaffold produced from chondroitin 6 sulfate, hyaluronan, and gelatin has been used to study cartilage dedifferentiation and synthesis of extracellular matrix to fill the cartilage lesions (187).

Inclusion of chondrotin sulfate in scaffold may promote the secretion of proteoglycan and type II collagen (188). Hyaluronan is present in high amounts in the ECM during embryonic cartilage development, and also facilitates the integration of engineered cartilage (189). Since gelatin is basically denatured collagen and presumably retains informational signals for cell attachment, it was hypothesized that a tri-copolymer formed from gelatin, chondrotin, and hyaluronan would mimic cartilage matrix and provide the necessary information for cell attachment to meet the requirement for dynamic reciprocity (the mechanism by which cell synthesizes and secretes ECM and is then, in turn, regulated by the ECM) (190) for cartilage tissue engineering. However the results of the study have shown that although chondrocytes were evenly distributes in the scaffold, secrete ECM, retain their phenotype, and secrete type II collagen but the amount of extracellular matrix is less to synthesize cartilage. The less than satisfactory amount of matrix production may be due to inhibition of cell mitosis by chondrotin. Further research in this field is under progress to stimulate cell proliferation in engineered tissue. The gelatin/chondrotin sulfate/hyaluronan tri-copolymer deserves to be tested as a candidate for cartilage tissue engineering. More ECM secretion with lacuna formation and more GAG production might be expected after longer periods of culture of chondrocytes.

Chitosan

Chitosan is a partially deacetylated derivative of chitin, found in arthropod exoskeletons. Structurally, chitosan is a linear polysaccharide consisting of β(1-4) linked D-glucosamine residues with a variable number of randomly located *N*-acetyl-glucosamine groups. It shares some characteristics with various glycosaminoglycans (GAGs) and hyaluronic acid present in articular cartilage. Depending on the source and preparation procedure, chitosan's average molecular weight may range from 50 to 1000 kDa.

Much of the potential of chitosan as a biomaterial stems from its cationic nature and high charge density in solution. Due to this it can form insoluble ionic complexes or complex coacervates with a wide variety of water-soluble anionic polymers like natural polysaccharides GAGs and alginate, as well as synthetic polyanions such as poly (acrylic acid) (191-193). Chitosan has been used as a biomaterial in drug delivery systems and it has recently drawn a considerable attention in its applications for repair of articular cartilage. Lu *et al.* (194) has demonstrated that the chitosan solution caused a significant increase in the density of chondrocyte in the knee articular cartilage, suggesting that the chitosan could be potentially beneficial to the wound healing of articular cartilage. Also Mattioli-Belmonte (195) has shown that the bone morphongenetic protein (BMP)-7, associated with *N*, *N*-dicarboxymethyl chitosan facilitates the repair of articular cartilage lesions. Chitosan forms insoluble ionic complexes with GAGs. A study has demonstrated that when associated with various polyelectrolytes GAGs chitosan has a protective effect against GAGs hydrolysis by their specific enzymes (193). These hydrogels have also been successfully used to deliver growth factors to chondrocytes (196).

Further studies are necessary to investigate the material and mechanical properties (such as porosity, compressive elastic modulus, permeability, and viscoelastic properties) of the chitosan scaffold. A development of a chitosan-based material that can support chondrogenesis may be significant not only in terms of the quality of neocartilage produced, but also in terms of the ability of that tissue to integrate with the host matrix.

Synthetic Scaffold Materials

Polylactic Acid (PLA)

The most widely used synthetic polymer to construct temporary scaffolds for tissue engineering includes poly (glycolic acid) (PGA), poly (lactic acid) (PLA), and their copolymer poly (D-L-lactic-co-glycolic acid) (PLGA) (118, 197-199). PLA, PGA and PLGA are biocompatible and are among few synthetic polymers approved by the Food and Drug Administration (FDA) for certain human clinical application such as surgical sutures and some implant devices.

Chu *et al.* (200) performed a study in rabbits in which allogenic perichondral cell/PLA composites were implanted for repair of cartilage and in same type of study, Dounchis *et al.* (201) used autogenous perichondrial cells. The results of both the studies showed grossly successful repairs followed by inconsistent subchondral bone reformation. Also the newly formed cartilaginous tissue had biochemical properties inferior to those of normal articular cartilage. More recently, Giurea *et al.* (202) modified the procedure by pre incubating the cells in matrices prior to implantation for improving their retention in scaffold. Caterson *et al.* (203) have shown that marrow MSCs displayed chondrocytic differentiation when cultured within PLA or PLA/alginate scaffolds in presence of growth factors TGFβ-1. Goomer *et al.* (204) used a PLA scaffold to implant perichondral cells transfected with parathyroid hormone receptor TGF β-1 gene in rabbits. In a composite scaffold study Frenkel *et al.* (205) used a D,D,L,L-polylactic acid and collagen device that has separate tissue-specific microenvironments for regeneration of bone and cartilage. The treatment of BMP-2 induced a high-quality, hyaline tissue that integrated well with the host tissue for 24 weeks.

Polyglycolic Acid (PGA) and Copolymers

Chondrocytes from human, bovina and many other sources have been cultured on the copolymers of poly glycolic acid with poly lactic acid (PLA), poly ethylene glycol and other viscous materials. These copolymers have been widely used for over 20 years. (206).

Freed (118) conducted a study to demonstrate that *in vitro* chondrocytes grown on PGA and PLLA copolymer for 1-6 months maintained the 3D shapes of the original polymer scaffolds, appeared glistening white macroscopically, contained S-GAG and type II collagen and histologically resembled to cartilage. The chondrocytes secreted cartilage matrix to fill the void spaces in polymer scaffolds that simultaneously biodegrade. This study demonstrated the feasibility of culturing isolated chondrocytes on biodegradable polymer scaffold to regenerate 3 D neocartilage.

Utilization of PLLA/PGA copolymer poly (DL-lactic-co-glycolic acid) (PLGA) allows one to control the rate of degradation of the scaffold. It is significant as residence time of the polymer at the defect site must be sufficient to serve its scaffolding purpose- without being so long as to impede tissue degeneration. Recently, Cohen *et al.* (207) used alginate beads to deliver chondrocytes homogeneously throughout PLA/PGA pads implanted in rabbits; at 12 weeks they demonstrated good histological and biochemical response. Sherwood *et al.* (208) designed a biphasic scaffold an upper, cartilage region composed of D, L-PLGA/L-PLA and a lower, bone portion consisting of an L-PLGA/tricalcium phosphate composite. Chondrocytes preferentially attached to the cartilage portion of the device, and biochemical and histological analyses showed that cartilage formed during a 6-week *in vitro* culture period. In another study, Kawanishi *et al.* (209) utilized only PLGA copolymer to develop a bilayered porous scaffold that combine freeze-drying and porogen leaching techniques. Thus pore size at both the layers can be precisely controlled. These bilayered porous scaffolds were revealed to have good potential in generating artificial cartilage.

Polyethylene Glycol (PEG) Hydrogel Scaffold

Hydrogels are attractive material for cartilage engineering because of their high water retaining properties and tissue like elastic properties. Hydrogels provide an environment for chondrocytes to retain their native form surrounded by lacunae. Examples of naturally occurring hydrogels are alginate and fibrin glue, which have been extensively used for encapsulation of chondrocytes and successfully regenerate cartilaginous tissue. However, limitations exist with naturally occurring hydrogels with respect to gel properties, gelation processes and also mechanical compressive properties. As an alternative to these natural hydrogels, synthetic and photocrosslinkable hydrogels based on polyethylene glycol (PEG) have been developed. (210). These hydrogls provide greater control over the hydrogel macroscopic properties. In addition, photo polymerization of hydrogles allow temporal and spatial control of the polymerization and gelation process, which increase the ease of handling for the surgeons as well as the ability to form complex architecture *in vivo*.

Further research is in progress to determine optimum crosslinking density in PEG hydrogels such that the gel mechanics is maximized, while a critical water content for cell viability and critical mesh size for extracellular matrix distribution is maintained.

Polyurethanes Scaffold

For years, biomedical segmented polyurethanes have extensively been used in various implantable devices like pacing leads insulation, indwelling catheters, intra-aortic balloons and mammary implants (211,212). While one of the primary features of these polyurethanes is their molecular stability *in vivo*, for the last several years there has been an increasing interest in designing biodegradable polyurethanes, which degrade *in vivo* to nontoxic by-products (213-216). Recently biodegradable polyurethanes have successfully been used as cancellous bone graft substitutes in animals. (217,218) However, very little attention has been paid to biodegradable polyurethanes as scaffolding materials for chondrocytes. To date, only hydroxybutyrate-co-valerate polyol-based polyurethane has been used to culture rat chondrocytes (219). A recent study by Grad (220) demonstrated that three-dimensional porous scaffolds prepared from linear biodegradable polyurethanes might serve as a carrier for cartilage tissue engineering, although several limitations need yet to be overcome. Firstly, the retention of matrix macromolecules in the scaffold has to be improved, and secondly, the scaffold structure should be modified to maintain the appropriate chondrocytic phenotype of the cells. A possible strategy to overcome these problems might be to decrease the scaffold pore size without impairing a homogeneous cell distribution within the scaffold. Also a too low pore size in the scaffolds may potentially limit the access to nutrients (111). Alternatively, the cells might be suspended in a gel matrix, such as agarose, alginate, chitosan, fibrin, or collagen, in order to improve extracellular matrix retention, before seeding them into the scaffold. The scaffold would have important function in helping to maintain the mechanical rigidity while the gel matrix would permit, in addition to improving the retention rate, a three-dimensional configuration of the chondrocytes that would facilitate the preservation of their differentiated phenotype (169).

Poly (hydroxybutyrate-co-hydroxyhexanoate) Scaffolds

Polyhydroxybutyrate (PHB) is finding some interesting potential applications as an implant material due to its biocompatibility and resorbability. PHB has already been used as temporary stents, bone plates, patches, nails and screws. The limitation associated with its use in soft tissue engineering is its brittle mechanical properties. To circumvent this, materials such as triethylcitrate have been used as plasticizers with PHB. Blend copolymer of PHA consisting of 3-hydroxybutyrate and 3-hydroxyvalerate (P (HB-co-HV)) were reported to be widely used with its improved mechanical properties over that of PHB (221). PHA with better elastomeric properties, such as poly (hydroxybutyrate-co-hydroxyhexanoate) (PHBHHx), is a potential candidate to meet special application as scaffolds in cartilage engineering (222) and research are being carried out in this direction.

Study by Deng (223) suggests the production of extracellular matrix of rabbit articular cartilage chondrocytes grown on scaffolds of PHB blended with PHBHHx. The m-RNA level of type II collagen of chondrocytes seeded on scaffolds consisting of PHBHHx were higher than that of PHB-only scaffold, throughout the culture period, suggesting the positive effect of PHBHHx on extracellular matrix production. Second-harmonic generation (SHG) imaging technique, combined with confocal fluorescence microscopy (CFM) revealed that PHBHHx in PHB scaffold provided better surface properties for anchoring type II collagen filaments and their penetration into internal layers of the scaffolds. Also glycosaminoglycan showed a sharp increase in construct of PHB/PHBHHx scaffold after 7-day cultivation, while only a small increase was observed in all other tested scaffolds. It appears that the presence of right proportion of PHBHHx in the composite system of PHB/PHBHHx, highly favors the production of extracellular matrix of articu

Titanium Alloy and Calcium Polyphosphate Scaffolds

One of the procedures for cartilage repair involves transplantation of osteochondral plugs (subchondral bone and overlying articular cartilage) obtained from donor sites within the same joint, into the cartilage defect (224, 225). This treatment is termed as 'mosaicplasty' and has been widely used over the past several years. But there are several major disadvantages with this technique like donor site morbidity, movement of implanted plugs relative to each other, the formation of repair tissue between plugs (*i.e.* spaces between juxtaposed round plugs), and the unpredictable integration of transplanted plugs with host tissues (226, 227). Thus misalignment of the grafts during placement can compromise graft viability because of formation of fibrous tissue at the perimeter of the grafts and/ or abrasion of graft cartilage if protruding above the intended surface of articulation.

To circumvent some of these problems an approach has been developed that consists of generating a biphasic (*i.e.* cartilage and porous bone substitute) "osteochondral-type" plug in vitro (228, 229). This approach precludes the need to harvest donor plugs. The biphasic plug is composed of cartilaginous tissue that is formed by isolated chondrocytes placed on the surface of the porous substrate. The porous substrates, which were utilized, are porous calcium polyphosphate (CPP) and porous titanium alloy (Ti6Al4 V) (229, 230). Although these substrates have very different compositions and surface roughness, chondrocytes cultured on the top surface of these materials will form a continuous layer of cartilaginous tissue. The cells maintain their phenotype as characterized by the synthesis of large proteoglycans and type II collagen. Chondrocytes attached to 3D-porous substrates, Ti alloy and CPP, remained less spread than cells in monolayer culture and exhibited a ring-like, subcortical organization of actin filaments. The presence of these features may be involved in favoring tissue formation by the cells. Further studies are ongoing to evaluate how these porous materials induce these effects in chondrocytes. Knowing this it should be possible ultimately to synthesize *in vitro* the novel materials that would incorporate the characteristics of osteochondro graft and enhance cartilage tissue formation.

PLLA-Collagen Hybrid Scaffold

To integrate into the skeletal system, adequate mechanical strength is required of tissue engineered cartilage grafts to enable surgical handling and mechanical loading. Because biodegradable synthetic polymers can easily be processed into intended shapes and have good mechanical strength, resulting in easy handling and withstanding tissue collapse, they have been used as scaffolds for tissue engineering of cartilage. But lack of cell interaction and the hydrophobic property of these materials hinder uniform cell seeding. Many efforts have been made on seeding chondrocytes on synthetic polymer scaffolds but it is still difficult to achieve uniform cell distribution. The concept of hybrid scaffolds is a promising one where synthetic material is hybridized as skeletons with other more mechanically weak but bioactive natural materials. In few recent studies PLLA sponge of large pore size, hybridized with collagen sponge has been used as the scaffold for *in vivo* tissue engineering of bovine articular cartilage (231-234). The PLLA–collagen sponge retained the advantages of both PLLA and collagen, and facilitated the formation of cartilaginous tissue, suggesting that it would be a useful scaffold for cartilage tissue engineering. Further quantitative studies are required to address the precise seeding efficiency utilized by the hybrid scaffold. The effect on biomechanical composition and mechanical character of constructs should also be evaluated.

Table 1.2: Summary of Performance of Various Scaffolds in Cartilage Engineering

Scaffold	*Advantages*	*Disadvantages*	*References*
Heterologous	1. Native protein of body	1. Incomplete integration with host tissue	149, 150, 165, 172, 235
A. Natural Scaffolds	2. Good mechanical properties	2. Xerogenic source may provoke immune response	
(i) Collagen	3. Can form various shapes	3. Early repair may thin over time	
	4. Excellent histological results when carrying BMP-2 and growth factors		
(ii) Fibrin	1. Can be obtained from own blood	1. Poor mechanical properties	173, 178, 179, 180
	2. Provide good cell attachment and wound healing	2. Does not permit host cell in growth	
	3. Produce good result as carrier of c-DNA for growth factor and BMP-2		
(iii) Agarose/alginate	1. Uniform cell distribution within implant	1. Foreign body giant cell reaction (in agarose)	181, 182, 183, 236
	2. Injectible and easily degradable	2. Inferior mechanical properties	
	3. Good histological and biochemical properties	3. Does not resorb well	
(iv) Hyaluronan	1. Good binding with joints	1. Poor biochemical and mechanical assessment	138, 184, 185, 237
	2. Native to joints	2. May induce chondrolysis	
	3. Binds to many proteins		
(v) Chitosan	1. Binds to polyanionic compounds to form hydrogels	1. Poor mechanical properties	191, 192, 195
	2. Deliver growth factors and BMP-7 along with chondrocytes	2. Inferior biochemical properties	
		3. May induce foreign body response	

Contd...

Table 1.2–Contd...

Scaffold	*Advantages*	*Disadvantages*	*References*
B. Synthetic Scaffolds	1. Biocompatible and biodegradable	1. Inferior biochemical properties	200, 201, 202, 205
	2. Good interaction with host in presence of BMP-2	2. Gross repair with inconsistent sub-chondral bone formation	
(i) Poly lactic acid (PLA)	1. Excellent retention at implant site		
(ii) Poly glycolic acid (PGA)	1. Good short term results in some animals		118, 206
(iii) PLGA Copolymer	1. Controlled pore size		207, 208, 209
	2. Good histological and biochemical results		
	3. Maintain 3D shape		
(iv) Polyethylene glycol (PEG)	1. High water retaining properties	1. Poor Biomechanical properties	210
	2. Controllable pore size and elasticity		
Autologous Scaffolds			
(i) Perichondrium/ Periosteum	1. Applicable for young patients with part of healthy cartilage	1. Incomplete filling of defects	130,196
	2. Ease of surgical handling	2. Not satisfactory with people age 40 years	
	3. Carry undifferentiated progenitor cell	3. Graft detachment during movement	
(ii) Mosaicplasty	1. Pain relief and restoration of joint function	1. Creates defects in non-weight bearing regions	131, 225
(iii) Autologous chondrocytes transplants	1. Hyaline like repaired tissue	1. Requires double surgery (harvesting /implantation)	176, 238
	2. Graft matures with time	2. Technically difficult surgery	
	3. Pain relief and recovery of function	3. Cells are expanded in vitro	

The designing and evaluation of scaffolds is one of the major upcoming field of tissue engineering and several materials are in the infancy of investigation for appropriateness for cartilage repair, including poly (ethylene glycol)-terephthalate and poly (butylenes terephthalate) copolymer blocks, oligo (poly (ethylene glycol) fumarate, poly (*N*-isopropylacrylamide), and carbon fiber scaffolds. The versatility of many of the scaffolds discussed above is often counter balanced by a lack of one or more of the cited parameters of successful implant design. The vast effort being mounted to formulate scaffolds with multifarious role in tissue-regeneration will result in the highly efficient clinical devices for articular cartilage repair.

Critical Conditions in Articular Cartilage Tissue Engineering

Some of the critical aspects, which have to be monitored closely during *ex vivo* cartilage engineering are as following:

Role of Mechanical Stimulations

It is clear that mechanical stimuli have their role in synthesis and organization of components of articular cartilage that in turn affect the cellular microenvironment. *In vitro* studies have demonstrated that like most cell types, chondrocytes are also connected to their extracellular environment via focal adhesions, which mediate mechanotransduction and affect fundamental cellular activities. A scaffold coated with any of the adhesion molecule like fibronectin and laminin may serve as a substratum to promote cell attachment and tissue formation.

The influence of mechanical stimuli on the development of articular cartilage is especially important because other stimuli provided via vascular, lymphatic and neural systems are not necessarily available to chondrocytes due to avascular, aneural and alymphatic nature of cartilage. Application of mechanical stimuli like hydrostatic pressure, fluid-flow-induced shear, or interaction with a solid object generally promotes expression of cartilage molecules and synthesis of a cartilaginous material. (239,240). Although the precise signaling pathways that are involved in mediation of mechanostimuli are not completely understood, evidence suggests that some types of mechanical stimuli like turbulent flow, or suspended as freely floating composites in rotating vessels, are desirable for articular cartilage synthesis and modeling and clear differences were observed in the distribution of deposited matrix within the PGA scaffolds and in properties of the synthesized cartilaginous material (241). The metabolic activities of chondrocytes are significantly affected by varying load parameters, frequency and load duration. The role of mechanical stimuli would be very important while designing the bioreactor for cartilage genesis *in vitro.*

Role of Cell Seeding Density

In the literature, few studies have addressed the effects of the initial cell density on cartilage genesis. In a recent study Mauck (242) investigated the role of initial cell seeding density and nutrient supply on cartilage formation in agarose hydrogels and studied the effect of mechanical intermittent deformational loading. The young's modulus and dynamic modulus were calculated and showed that scaffolds seeded at high cell density (60 million cells/ml) show >2 fold increase in material properties in comparison to the constructs with low cell density (10 million-cells/ml). Also the chondrocytes seeded at density less than 10 million cells per ml at polylactide–polyglycolide scaffolds results in very little cartilaginous material synthesis (243). As the cell seeding density increases, the integrity of the cartilaginous product improves and its distribution becomes more homogeneous. Another systemic study investigated the effect of initial cell seeding density on cartilage formation and revealed that seeding density ranging from 20-100 million cells per ml resulted in formation of clinically appropriate cartilage when implanted subcutaneously into nude mice; whereas lower seeding density (*e.g.*, 10 million cells per ml) exhibited reduced cartilaginous formation as assessed by wet weight and thickness (244).

Role of Media Composition

To obtain the optimum number of chondrocytes and their secreted matrix products is the core of *in vitro* cartilage genesis. A number of different growth media with varying nutrient compositions are commercially available. The medium for chondrocyte culture should contain an energy source, inorganic salts, amino acids, vitamins and a buffering component, usually either a bicarbonate buffer or HEPES. Glucose, pyruvate and glutamate are included in growth medium as source of energy. Fetal bovine serum is a complex undefined media containing mitogenic peptide growth factors and many attachment factors to maintain differentiation and viability of chondrocytes. Other factors, which may be included, are those, which promote cell adhesion and spreading (*e.g.*, fibronectin and vitronectin), trace minerals, transport proteins such as transferrin and albumin, growth factors, and hormones. Typically 10% of the volume of growth media consists of fetal bovine serum or calf serum. For the synthesis of cartilage, supplements like ascorbic acid, proline, and some non-essential amino acids may be included. The chondrocyte culture is maintained at 37°C in presence of 5% CO_2 and penicillin and streptomycin antibiotics. A combination of commercially available media components like Dulbecco's Modified Eagle Medium (DMEM) can easily supply all these components. Various growth factors like TGF-β, IGF-I, b FGF and BMPs may also be added in the medium to enhance the synthetic properties of cartilage.

Role of Scaffold Surface Properties and their Modifications

There are three mechanisms for attachment of cells to scaffolds:

1. Coating with physiological adhesion proteins such as fibronectin (245),
2. Cell contact with specific synthetic peptides derived from extracellular matrix molecules (243, 246, 247) and
3. Cell interaction with a suitably charged substratum such as a polylysine-coated surface.

The serum components like fibronectin and vitronectin are primary glycoproteins and promote adhesion of various synthetic peptides that are known to interact with cell surface adhesion molecules (249). Use of polylysine allows cells to spread on substratum via charge-mediated interactions (243). Also synthetic peptides, which interact with cell surface adhesion molecules like integrins (246) and proteoglycans (248) are likely to play a significant role in cartilage engineering.

Challenges for Tissue Engineered Cartilage

Cartilage tissue engineering has emerged as a potential alternative to joint replacement and represents an exciting direction in the efforts to solve the problem of cartilage regeneration. The aspect of tissue engineering that involves *ex vivo* fabrication of cartilage constructs occupies a central position in addressing this complex problem in a logical manner. Functional restoration of cartilage tissue is highly desirable yet remain medically challenging.

Mechanical Stability of Engineered Cartilage

The primary goal of *in vitro* engineering hyaline articular cartilage is to achieve the native biomechanical properties of the tissue. Although various studies have suggested as described in above sections that *in vivo* tissue-engineered cartilage achieves some pressure resistance and stiffness values which is comparable to those of native human cartilage. However, yet it is not clear that up to what extent of physiological loading the newly construct cartilage transplants could remain resilient. Also the regenerated cartilage reveals clear differences between native and transplanted cartilage as the distribution of cells lack typical zonal organization and appears somewhat random. Also the typical collagen architecture gets altered in more loosened and irregular helix formation. The water retaining property of the tissue is also under investigation, which is a very important feature of normal cartilage for its weight bearing compressive stiffness properties.

Fixation of Graft at the Site of Injury

Another major problem in cartilage engineering is the fixation of the cartilage transplant to the subchondral bone at the site of injury. In theory, the artificially grown cartilage layers could be attached directly to the defect joint surface using fibrin glue or collagen-fibrin layer. Also the use of resorbable pins has been mentioned in various studies (250, 251). Currently, clinical studies on patients with *in vitro* engineered cartilage flaps are ongoing. Another strategy that has gained attention is to develop an osseo-integrating interface for cartilage implant using either a calcium carbonate or hydroxyapatite–tricalcium phosphate (HA–TCP) scaffold or an engineered osteoblast layer (252). The ultimate aim of these studies is to achieve a permanent, solid connection between cartilage and bone tissue.

Immunological Aspects

Cartilage was long considered an immunologically privileged tissue. However, studies have shown that cartilage, and specifically chondrocytes, do possess antigens and are capable of eliciting an immune response. Two major immunological risks are linked to the use of engineered cartilage transplants. First, the invasion of scaffold by host giant cells or granulocytes. Second, the exposures of hidden cell surface and matrix protein epitopes, which are usually masked from the immune system and might therefore, generate immune response. The persons who have been operated for cartilage transplants, showed humoral reactions against type IX and XI collagens, which are associated with collagen, type II fibrils (253). In response to isolated chondrocyte allografts in full-thickness articular cartilage defects in rabbit knees, host lymphocytes accumulated around the allografts. According to the study by Kawabe (178) only 26% of the grafts contained viable cartilage cells after three weeks and both cell-mediated cytolytic reaction and a humoral immune response were present. Collagen and proteoglycan degrading enzymes were secreted in response to a factor produced by macrophases; and their further synthesis is inhibited by lymphocytes produced factors. Other studies have shown that antibodies may also be directed against other antigenic sites on chondrocytes or against specific sites in the surrounding matrix (254). Many other reports have confirmed the antigenicity of isolated chondrocytes; however, the immune response does not cross the intact matrix (255, 256). The matrix appears to prevent recognition of antigens on the chondrocytes, protecting them from the host immune response (257, 258)

Clearly, even though tissue engineering is usually regarded as an autologous therapy, a major clinical breakthrough would be, to solve the immunological problems associated with the technique.

Future Directions in Cartilage Engineering

The key to successful repair and regeneration of cartilage is to provide the repair site with sufficient chondrogenic cells in a suitable delivery vehicle to ensure maximal differentiation and deposition of right extracellular matrix. New optimized culture methodologies and bioreactors that provide appropriate mechanical and other guidance clues in order to ensure the successful and long-term functioning of engineered tissue. As we gain more and more information about the identity of morphogens for chondrocyte differentiation, it might be possible to orchestrate massive cartilage regeneration by clever combination of smart 3D scaffolds and such morphogenic factors. The management of cells both *in situ* and *ex vivo*, will be crucial to the success of such tissue engineering efforts. Also concentrated efforts are on to optimize the use of biochemical regulatory signals and growth factors like TGF-β, recombinant human BMP-2 (rhBMP-2) in chondrocytes differentiation which may help in understanding the complex phenomenon of cartilage degeneration. In the future, it can be envisioned that a combination of growth factors or cytokines may be incorporated into scaffolds to modify different cell functions, such as differentiation, metabolism, and synthesis of extracellular matrix. Recently, researchers have begun to combine the above chondrocyte culture models with external mechanical stimuli so as to improve the quality of generated cartilage tissue. Regulation of the biomechanical environment is a powerful modulator in tissue engineering of articular cartilage. The application of extrinsic mechanical forces can have a profound effect on chondrocyte function and cartilage regeneration. Effects of agitation/stirring, microgravity, and shear stress have a direct role on cell metabolism and extracellular matrix synthesis. It is essential to quantify the stress–strain fields experienced by individual cells attached on scaffolds and how these mechanical stimuli affect their mechanotransductional abilities.

Modification of scaffold surface material properties in order to enhance cell adhesion and to promote repair of partial and full thickness defects of articular cartilage are only now beginning to be considered. In the future, it may be desirable to modify scaffold external and internal surfaces in a preferential way such that preferred chemotaxis can be achieved. Another promising avenue is the identification and characterization of novel cell adhesive extracellular matrix molecules that may play real roles in cartilage development.

A very important aspect of research in chondrocyte culture and artificial cartilage engineering is development of clinical standards for assessments of neocartilage. As may be expected when assessing complex biological systems, the variables, which potentially affects the experimental outcomes of cartilage repair, needs thorough evaluations. For example evaluation of scaffolds with regards to mechanical properties, permeability, porosity, material properties, degradation rate, and biocompatibility. The experimental approach employed in the *in vivo* assessment of engineered cartilage also needs to be standardized with regards to the animal models used, surgical approach, defect size, location, defect creation, and duration of healing. Standard protocols should be allowed for multidisciplinary assessment of healing using well-defined tests.

It is essential and desirable that multiprong and multidimensional efforts should be made sincerely for the development of clinical standard protocol for chondrocyte cell tissue engineering. So the gigantic problem of degenerative joint disease like osteoarthritis should be tackled and diagnosed at early stages of development and remedial measures for the treatment and repair process may be initiated for the prevention of this dreaded disease.

REFERENCES

1. Mankin, H.J. and Lippiello, L. (1969) The turnover of adult rabbit articular cartilage **J. Bone Jt. Surg.** (51A) 1591.
2. Stockwell, R.A. (1979) Biology of cartilage cells, Cambridge, Cambridge University press.
3. Muir, H. (1995) The chondrocyte, architect of cartilage: Biomechanics, structure, function and molecular biology of cartilage matrix macromolecules. **Bioessays** (17) 1039-1048.
4. Mayre, R. Collagens: (1989) What is their function, are they involved in articular disease. **Arthritis Rheumatism (**32) 241-246.
5. Eyre, D.R., Wu, J.J. and Woods, P.E. (1991) The cartilage collagens: Structural and metabolic studies. **J. Rhematol**. (18) 49-51.
6. Roughley, P.J. and Lee, E.R. (1994) Cartilage proteoglycans: Structure and potential function. **Microsc. Res. Technol**. (28) 385-397.
7. Mankin, H.J., Dorfman, H., Lippiello, L. and Zarins, A. (1971) Biochemical and metabolic abnormalities in articular cartilage from osteoarthritic human hips. **J. Bone Joint Surg.** (53 A) 523-537.
8. Maroudas, A. (1976) Balance between swelling pressure and collagen tension in normal and degetenerative cartilage. **Nature** (260) 808-809.

9. Imai, K., Ohta, S., Mastumoto, T., Fujimoto, N., Sato, H. and Seiki, M. (1997). Expression of membrane type I matrix metalloproteases and activation of progelatinase A in human osteoarthritic cartilage. **Am. J. Pathol.** (151) 245-256.

10. Ohta, S., Imai, K., Yamashita, K., Mastumoto, T., Azumano, I. and Okada, Y. (1998) Expression of matrix metalloproteases-7 (matrilysin) in human osteoarthritic cartilage. **Lab. Investi. (**78) 79-87.

11. Lohmander, L.S., Neame, P.J. and Sandy, J.D. (1993) The structure of aggrecan fragments in human synovial fluid-evidence that aggrecanase mediate cartilage degeneration in inflammatory joint disease, joint injury and osteoarthritis. **Arthritis Rheum**. (36) 1214-1222.

12. Mynarik, V. and Trating, S. (2000) Physiochemical properties of normal articular cartilage and its MR appearance. **Invest. Radiol**. (35) 589-594.

13. Hauselmann, H.J., Masuda, K., Hunziker, E.B., Neidhart, M., Mok, S.S., Michel, B.A. and Thonar, E.J.M.A. (1996) Adult human articular chondrocytes cultured in alginate gel forms a matrix similar to native human articular cartilage. **Am. J. Physiol**. (Cell Physiol.) (271) 742-742.

14. Aydellotte, B.A., Schleyerbach, R., Zeck, B.J. and Kuettner, K.F. (1986) Articular chondrocytes cultured in agarose gel for study of chondrocytic chondrolysis. In *Articulur cartilage biochemistry* (eds. Kuettner, K.E., Schleyerbach R. and Hascall V.) Ravan press, New York. pp235-254.

15. Adolphe, M. and Benya, P. (1992) Different types of cultured chondrocytes- the in vitro approach to the study of biological regulation. In *'Biological regulation of chondrocytes'* (ed M. Adolphe), CRC press Boca Raton, pp105-139.

16. Hauselmann, H.J., Fernandes, R.J., Mok, S.S., Schmid, T.M., Block, M.B., Kuettner, K.E. and Thonar, E.J. (1994) Phenotypic stability of bovine articular chondrocytes after long-term culture in alginate beads. **J. Cell. Sci**. (107) 17-27.

17. Handley, C.J., McQuillan, D.J., Campbell, M.A. and Bolis, S. (1986) Steady state metabolism in cartilage explants. In *'Articulur cartilage biochemistry'* (eds Kuettner, K.E., Schleyerbach, R. and Hascall V.C.) Raven Press, New York, pp163-179.

18. Benya, P.D. and Shaffer, J.D. (1982) Dedifferentiated chondrocyte reexpress the differentiated collagen phenotype when cultured in agarose gel. **Cell** (30) 215-224.

19. Buschman, M.D., Gluzband, Y.A., Grodinzsky, A.J., Kimmura, J.H. and Hunziker, E. (1992) Chondrocyte in agarose culture synthesizes a mechanically functional extracellular matrix. **J. Orthop. Res.** (10) 745-758.

20. Guo, J., Jourdian, G.W. and Maccallumn, D.K. (1989) Culture and growth characteristics of chondrocytes encapsulated in alginate beads. **Connect. Tissue Res**. (19) 277-297.

21. Hauselmann, H.J., Aydellotte, M.B., Schumacher, B.I., Kuettner, K.E., Gitelis, S.H. and Thonar, E.J.M.A. (1992) Synthesis and turnover of proteoglycan by human and bovine articular chondrocytes cultured in alginate beads. **Matrix** (12) 116-129.

22. Bonaventure, J., Kadhom, N., Cohensolal, L., Ng., K.H., Bourguignon, J., Lasselin, C. and Frisinger, P. (1994) Reexpression of cartilage specific genes by dedifferentiated human articular chondrocytes cultured in alginate beads. **Exp. Cell Res**. (212) 99-104.

23. Kimmura, T., Yasui, N., Ohsawa, S. and Ono, K. (1984) Chondrocyte embedded in collagen gel maintains cartilage phenotype during long term cultures. **Clin Orhtop. Relat. Res**. (186) 231-239.

24. Maor, G., Vonder mark, K., Reddi, H., Heingard, D., Franzen, A. and Sibermann, M. (1987) Acceleration of cartilage and bone differentiation of collagenous substrata. **Collagen Relat. Res**. (7) 351-370.

25. Solursh, M. and Meir, S. (2001) Effects of cell density on the expression of differentiation by chick embryo chondrocytes. **J. Exp. Zool**. (187) 311-322.

26. Bywater, *E*.G.L. (1937) The metabolism of joint tissues. **J. Path. Bact**. (44) 237.

27. Poole, C.A., Flint, M.H. and Beaumont, B.W. (1987) Chondrons in cartilage: Ultrastructural analysis of the pericellular microenvironment in adult human articular cartilage. **J. Orthop. Res**. (5) 509-522.

28. Grodzinsky, A.J. (1983) Electromechanical and physicochemical regulation of cartilage strength and metabolism. **Crit. Rev. Bioengg**. (9) 133-199.

29. Maroudas, A. (1979) Physico-chemical properties of articular cartilage. In: *Adult articular cartilage*, ed. Freeman, M.A.R. (London: Pitman medical) pp215-290.

30. Holm, S., Maruodas, A., Urban, J., Selstam, G. and Nachemson, A. (1981) Nutrition of the intervertebral disc: Solute transport and metabolism. **Conn. Tiss. Res.** (8) 101-119.

31. Maroudas, A., Popper, O. and Grushko, G. (1992) Partition coefficient of IGF-1 between cartilage and external medium in the presence and absence of FCS. **Trans. Am. Orhtop. Res. Soc**. (16) 398.

32. Hodge, W.A., Fijan, R.S., Carlson, K.L., Burgess, R.G., Harris, W.H. and Mann, R.W. (1986) Contact pressure in the human hip joint measured *in vivo*. **Proc. Natl. Acad. Sci**. (83) 2879-2883.

33. Afoke, N.Y., Byers, P.D. and Hutton, W.C. (1987) Contact pressure in the human hip joint. **J. Bone Joint Surg. Br**. (69) 536-541.

34. Bergmann, G., Graichen, F. and Rohmann, A. (1993) Hip joint loading during walking and running measured in two patients. **J. Biomech**. (26) 969-990.

35. Eisenhart, R., Adam, C., Steinlechner, M., Muller-Gerbl, M. and Ecstein, F. (1999) Quantitative determination of joint in configurity and pressure distribution during stimulated gait and cartilage thickness in the human hip joint. **J. Orthop. Res**. (17) 532-539.

36. Rydell, N. (1965) Forces in the hip joint, Part II intravital measurement. In *Biomechanics and Related Bioengineering Topics.* ed. by Kennedy R.M. (Pergamom Press, Oxford) pp 351

37. Paul, J.P. (1965) Bioengineering studies of the force transmitted by joints. Part II: Engineering analysis. In *Biomechanics and Related Bioengineering Topics.* ed. by Kennedy, R.M. (Pergamom Press, Oxford) pp 369.

38. Helminen, H., Jurvelin, J., Kjiviranta, I., Paukkonen, K., Saamanen, A.M. and Tammi, M. (1987) Joint loading effects on articular cartilage: a historical review. In *Joint loading: biology and health of articular structures.* Eds. Helminen, H., Kjiviranta, I., Tammi, M., Saamanen, A.M. and Paukkonen, K., Jurvelin, J. (Bristol: John Wright) pp 1-46.

39. Pond, M.J. and Nuki, G. (1973) Experimentally induced osteoarthritis in dog. **Ann. Rheum. Dis**. (32), 387-388.

40. Altman, R.D., Tenenbaum, J., Latta, L., Riskin, W., Blanco, L.N. and Howell, D.S. (1984) Biomechanical and biochemical properties of dog cartilage in experimentally induced osteoarthritis. **Ann. Rheum. Dis**. (43) 83-90.

41. Wu, D.D., Burr, D.B., Boyd, R.D. and Radin, E.L. (1990) Bone and cartilage changes following experimental varus or vagus tibial angulation. **J. Orthop. Res**. (8) 572-585.

42. Bentley, G., Greer, R.B. (1971) Homotransplantation of isolated epiphyseal and articular cartilage chondrocytes into joint surfaces of rabbits. **Nature** (230) 184-197.

43. Brittberg, M., Lindhal, A., Nilsson, A., Ohisson, C., Isaksson, O. and Peterson, L. (1994) Treatment of deep cartilage defect in the knee with autologous chondrocyte transplantation. **New Eng. J. Med.** (331) 889-895.

44. Paige, K.T., Cima, L.G., Yaremchuk, M.J., Schloo, B., Vacanti, J.P. and Vacanti, C.A. (1996) De novo cartilage generation using calcium alginate-chondrocytes constructs. **Plastic Reconstruct Surg.** (97) 168-178.

45. Kawamura, S., Wakitani, S., Kimura, T., Maeda, A., Calpan, A.I., Shino, K. and Ochi, T. (1998) Articular cartilage. Rabbit experiments with a collagen gel-biomatrix and chondrocytes cultured in it. **Acta Orthop. Scand** (69) 56-62.

46. Pittenger, M.F., Mackay, A.M., Beck, S.C., Jaiswal, R.K., Douglas, R., Mosca, J.D., Moorman, M.A., Simonetti, D.W., Craig, S. and Marshak, D.R. (1999) Multilineage potential of adult human mesenchymal stem cells. **Science** (284) 143-147.

47. Thomson, J.A., Itskovitz-Eldor, J., Shapiro, S.S., Waknitz, M.A., Swiergiel, J.J., Marshall, V.S. and Jones, J.M. (1998) Embryonic stem cell lines derived from human blastocysts. **Science** (282) 1145-1147.

48. Wong, M. (2003) Chondrocyte in cartilage tissue engineering. **Bioforum Eur**. (7): 6, 324-325.

49. Grogan, S.P., Rieser, F., Winkelmann, V., Berardi, S. and Malini–Varlet, P. (2003) A static, closed and scaffold-free bioreactor system that permits chondrogenesis *in vitro*. **Osteoarthritis Cartilage** (11) 403-411.

50. Parkinnen, J.J., Ikonen, J., Lammi, M.J., Laakkonen, J., Tammi, M. and Helminen, H.J. (1993) Effect of cyclic hydrostatic pressure on proteoglycan synthesis in cultured chondrocytes and articular cartilage explants. **Arch. Biochem. Biophys** (300) 458-465.

51. Kim, Y.J., Sah, R.L., Grodinzsky, A.J., Plass, A.H. and Sandy, J.D. (1994) Mechanical regulation of cartilage biosynthetic behaviour: physical stimuli. **Arch. Biochem. Biophys** (311) 1-12.

52. Buschmann, M.D., Kim, Y.J., Wong, M., Frank, E., Hunziker, E.B. and Grodinzsky, A.J. (1999) Stimulation of aggrecan synthesis in cartilage explants by cyclic loading is localized to regions of high interstitial fluid flow. **Arch. Biochem. Biophys**. (366) 1-7.

53. Bonnassar, L.J., Grodinzsky, A.H., Frank, E.H., Davila, S.G., Bhaktav, N.R. and Trippel, S.B. (2001) The effect of dynamic compression on the response of articular cartilage to insulin like growth factor-1. **J. Orthop. Res**. (19) 11-17.

54. Jin, M., Frank, E.H., Quinn, T.M., Hunziker, E.B. and Grodinzsky, A.J. (2001) Tissue shear deformation stimulates proteoglycan and protein biosynthesis in bovine cartilage explants. **Arch. Biochem. Biophys (**395) 41-48.

55. Handley, C.J., Mcquillan, D.J., Campbell, M.A. and Bolis, S. (1986) Steady state metabolism in cartilage explants eds. Keuttner, K.E., Schleyerbach, R. and Hascall, V.C. (New York, Raven press), pp163-179.

56. Campbell, N.A., Handley, C.J., Hascall, V.C., Campbell, R.A. and Lowther, D.A. (1984) Turnover of proteoglycan in culture of bovine articular cartilage. **Arch. Biochem. Biophys**. (234) 275-281.

57. Hascall, V.C., Morales, T.I., Hascall, G.K., Handley, C.J. and McQuillan, D. (1983) Biosynthesis and turnover of proteoglycan in organ culture of bovine articular cartilage. **J. Rheumatol. Suppl**. (10) 45-52.

58. Morales, T., Wahl, L.M. and Hascall, V.C. (1984) The effect of lipolpolysaccharides on biosynthesis of proteoglycans from calf articular cartilage cultures. **J. Biol. Chem**. (259) 6720-6729.

59. Bartholomew, J.S., Handley, C.J. and Lowther, D.A. (1985) The effect of trypsin treatment on proteoglycan biosynthesis by bovine articular cartilage. **Biochem. J**. (227) 429-437.

60. Tyler, J.A. (1985) Chondrocytes mediated depletion of articular cartilage proteoglycan *in vitro*. **Biochem. J**. (225) 493-507.

61. Sandy, J.D., Neame, P.J., Boynton, R.E. and Flannery, C.R. (1991) Catabolism of aggrecan in cartilage explants. **J. Biol. Chem**. (266) 8683-8685.

62. Tian, X., Chen, S., Morales, T.I. and Hascall, V.C. (1989) Biochemical and morphological study of steady state and lipopolysaccharide treated bovine articular cartilage explant cultures. **Connect. Tissue. Res. (**19) 195-218.

63. Morales, T.I. and Hascall, V.C. ('1989) Factors involved in the regulation of proteoglycan metabolism in articular cartilage. **Arthritis Rheum**. (32): 10, 1197-1201.

64. McQuillan, D.J., Handley, C.J., Campbell, M.A., Bolis, S., Milway, V.E. and Herington, A.C. (1986) Stimulation of proteoglycan biosynthesis by serum and insulin like growth factor-1 in culture bovine articular cartilage. **Biochem. J**. (240) 423-430.

65. Luyten, F.P., Hascall, V.C., Nissley, S.P., Morales, T.I. and Reddi, A.H. (1988) Insulin like growth factors maintain steady state metabolism of proteoglycan in bovine articular cartilage explants. **Arch. Biochem. Biophys**. (267) 416-425.

66. Roberts, A.B., Flanders, K.C., Kondaiah, P., Thompson, N.L., Van Obbereghen- Schilling, E., Wakefield, L., Rossi, P., de Drombrugghe, B., Heline, U. and Sporn, M.B. (1988) Transforming growth factor beta; biochemistry and roles in embryogenesis, tissue repair and remodeling and carcinogenesis. **Recent Prog. Horm. Res**. (44) 157-197.

67. Morales, T.I. and Hascall, V.C. (1989) Specificity of action of lipopolysaccaride on bovine articular cartilage metabolism: a comparison with interleukin-1. **Connect. Tissue Res**. (19) 255-275.

68. Weibkin, O.W., Hardingham, T.E. and Muir, H. (1975) Hyaluronic acid proteoglycan interaction and the influence of hyaluronic acid on proteoglycan synthesis by chondrocytes from adult cartilage. Extracellular matrix influences on gene expression. Edited by Slavkin, H., Greulich, R. (New York, Academic press).

69. Jurvelin, J., Kiviranta, I., Tammi, M. and Helminen, J.H. (1986) Softening of canine articular cartilage after immobilization of the knee joint. **Clin. Orthop. Rel. Res**. (207) 246–252.

70. Saamanen, A.M., Tammi, M., Kiviranta, I., Jurvelin, J. and Helminen, H.J. (1987) Maturation of proteoglycan matrix in articular cartilage under increased and decreased joint loading. A study in young rabbits. **Conn. Tissue Res**. (16) 63–175.

71. Behrens, F., Kraft, E.L. and Oegema, T.R. (1989) Biochemical changes in articular cartilage after joint immobilization by casting or external fixation. **J. Orthop. Res.** 7: 335–343, 1989.

72. Aydelotte, M.B., Greenhill, R.R. and Kuettner, K.E. (1988) Differences between subpopulations of cultured bovine articular chondrocytes II Proteoglycan metabolism. **Connect. Tissue. Res**.(18): 223-234.

73. Poole, C.A., Flint, M.H. and Beaumont, B.W. (1988 a) Chondrons extracted from canine tibial cartilage: Preliminary report on their isolation and structure. **J. Orthop. Res**. 6: 1988a

74. Graff, R.D., Kelley, S.S. and Lee G.M. (2003) Role of pericellular matrix in development of a mechanically functional neocartilage. **Biotechnol. Bioengg**. (82) 4, 457–464.

75. Benya, P.D. and Brown, P.D. Modulation of chondrocyte phenotype *in vitro*. (1985) In *Articular cartilage biochemistry;* Kuettner, K.E., Schleyerbach, R., Hascall, V.C. eds. New York, Raven Press pp 219-230,

76. Vondermark, K. (1986) Differentiation, modulation and dedifferentiation of chondrocytes. **Rheumatology** (10) 272-315.

77. Grundmann, K., Zimmermann, B., Barrach, H.J. and Merker, H.J. (1990) Behaviour of epiphyseal mouse chondrocyte populations in monolayer culture. **Virchows Arch. A. Path. Anat. Histol** (389) 167–187.

78. Quarto, R., Dozin, B., Tacchetti, C., Campanile, C., Malfatto, C. and Cancedda, R. (1990) *In vitro* development of hypertrophic chondrocytes starting from selected clones of dedifferentiated cells. **J Cell Biol.** (110) 1379–1386.

79. Shakibaei, M. (1995) Integrin expression on epiphyseal mouse chondrocytes in monolayer culture. **Histol. Histopathol.** (10) 339–349.

80. Benya, P.D., Padilla, S.R. and Nimni, M.E. (1977) The progeny of rabbit articular chondrocytes synthesize collagen types I, III and type I trimer, but not type II. Verifications by cyanogens bromide peptide analysis. **Biochemistry** (16) 865–872.

81. Vondermark, K., Gauss, V., Vonder, H. and Mu¨Ller, P. (1977) Relationship between cell shape and type of collagen synthesized as chondrocytes lose their cartilage phenotype in culture. **Nature** (267) 531–532.

82. Vondermark, K. (1980) Immunological studies on collagen type transition in chondrogenesis. **Curr. Top. Dev. Biol**. (14) 199–225.

83. Shakibaei, M., Zimmermann, B. and Merker, H.J. (1995) Changes in integrin expression during chondrogenesis *in vitro* (An immunomorphological study). **J. Histochem. Cytochem**. (43) 1061–1069.

84. Norby, D.P., Malemud, C.J. and Sokoloff, L. (1977) Difference in the collagen types synthesized by lapine articular chondrocytes in spinner and monolayer culture. **Arthritis Rheum**. (20) 709–715.

85. Schröter-Kermani, C., Hinz, N., Risse, P., Zimmermann, B. and Merker, H.J. (1991) The extracellular matrix in cartilage organoid culture: Biochemical, immunomorphological and electron microscopic studies. **Matrix** (11) 428–440.

86. Shakibaei, M., Schröter-Kermani, C. and Merker, H.J. (1993a) Matrix changes during long-term cultivation of cartilage (organoid or high-density culture). **Histol. Histopathol**. (8) 463–470.

87. Shakibaei, M., Abou-Rebyeh, H. and Merker, H.J. (1993b) Integrins in ageing cartilage tissue *in vitro*. **Histol. Histopathol**. (8) 715–723.

88. Gerstenfeld, L.C. and Landis, W.J. (1991) Gene expression and extracellular matrix ultrastructure of a mineralizing chondrocyte cell culture system. **J Cell Biol** (112) 501–513.

89. Zimmermann, B., Schröter-Kermani. C., Shakibaei, M. and Merker. H.J. (1992) Chondrogenesis, cartilage maturation and transformation of chondrocytes in organoid culture of mouse limb bud mesodermal cells. **Eur. Arch. Biol.** (103) 93–111.

90. Shakibaei, M. (2000) Adhesion rates and chondrocyte phenotype in monolayer cultures under the influence of collagen type II and IGF-I. **Acta Medica Iranica** (3) 138-147.

91. Thonar, E.J.M.A., Buckwalter, J.A. and Kuettner, K.E. (1986) Maturation related differences in the structure and composition of proteoglycan synthesized by chondrocytes from bovine articular cartilage. **J. Biol. Chem.** (261) 2467-2474.

92. Kimura, J.H., Hardingham, T.E. and Hascall, V.C. (1980) Assembly of newly synthesized proteoglycan and link proteins into aggregates in culture of chondrosarcoma chondrocytes. **J. Biol. Chem. (**255) 7134-7143.

93. Lohmander, L.S. (1990) Biosynthesis of cartilage proteoglycan: An analysis of posttranslational events by different *in vitro* labeling protocols. In '*Methods in Cartilage Research,* eds Maruodas, A. and Kuettner, K.E. (Academic press, San Diego, CA) pp148-152.

94. Hauselmann, H.J., Stefanovic-Racie, M., Michel, B.A. and Evans, C.H. (1998) Differences in nitric oxide production by superficial and deep human articular chondrocytes. Implications for proteoglycan turnover in inflammatory joint diseases. **J. Immunol**. (160) 1444-1448.

95. Chubinskaya, S., Huch, K., Mikecz, K., Cs- Szabo, G., Hasty, K.A., Kuettner, K.E. and Cole, A. (1996). Chondrocytes MMP-8: Upregulation of neutrophil collagenase by interleukin- 1b in human cartilage from knee and ankle joints. **Lab Invest**. (74) 232-240.

96. Brukner, P., Horler, I., Mendler, M., Houze, Y., Winterhalter, K.H., Eich-Bender, S.G. and Spycher, M.A. (1989) Induction and prevention of chondrocyte hypertrophy in culture. **J. Cell. Biol. (**109) 2537-2545.

97. Archer, C.W., Mc Dowell, J., Bayliss, M.T., Stephens, M.D. and Bentley, G. (1990) Phenotypic modulation in sub-population of human articular chondrocytes *in vitro*. **J. Cell Sci**. (97) 361-371.

98. Sun, D., Aydelotte, M.B., Maldonado, B., Keittner, K.E. and Kim, H. (1986) Clonal analysis of population of chondrocytes from rat chondrosarcoma in agarose culture. **J. Orthop. Res**. (4) 427-436.

99. Buschmann, M.D., Gluzband, Y.A., Grodinzsky, A.J. and Hunziker, E.B. (1995) Mechanical compression modulates matrix biosynthesis in chondrocyte/agarose culture. **J. Cell. Sci**. (108) 1497-1508.

100. Aydelotte, M.B., Schumacher, B.L. and Kuettner, K.E. (1992) Heterogeneity of articular chondrocytes. In *Articular Cartilage and Osteoarthritis* eds. Kuettner, K.E., Schleyerbach, R. Peyron, J.G., Hascall, V.C. New York: Raven press, pp-237-249,

101. Petit, B., Masuda, K., D'Souza, A.L., Otten. L., Pietryla, D., Hartmann. D.J., Morris. N.P., Uebelhart. D., Schmid, T.M. and Thonar. E.J.M.A. (1996) Characterization of crosslinked collagen synthesized by mature articular chondrocytes cultured in alginate beads: comparison of two distinct matrix compartments. **Exp. Cell. Res**. (255) 151-161.

102. Martinsen, A., Skjaak-Braek, G. and Smidsroed, O. (1989) Alginate as immobilization material: I Correlation between chemical and physical properties of alginate gel beads. **Biotechnol. Bioengg.** (33) 79–89.

103. Rowley, J.A., Madlambayan, G. and Mooney, D.J. (1999) Alginate hydrogels as synthetic extracellular matrix materials. **Biomaterials** (20) 45–53.

104. Mok, S.S., Masuda, K., Häuselmann. H.J., Aydelotte, M.B. and Thonar, E.J. (1994) Aggrecan synthesized by mature bovine chondrocytes suspended in alginate. Identification of two distinct metabolic matrix pools. **J. Biol. Chem.** (269) 33021–33027.

105. Kamada, H., Masuda, K., D'souza, A.L., Lenz, M.E., Pietryla, D., Otten, L. and Thonar E.J.M.A. (2002) Age related differences in the accumulation and size of hyaluronan in alginate culture. **Arch. Biochem. Biophys** (408) 192-199.

106. Genes, N.G., Rowley, J.A., Mooney, D.J. and Bonassar, L.J. (2004) Effect of substrate mechanics on chondrocyte adhesion to modified alginate surfaces. **Arch. Biochem. Biophys**.

107. Cimpean, A., Zarnescu, O., Caloianu, M., Moldovan, L. and Efimov, N. (2000) Glycosaminoglycan Synthesis by Human Osteoarthritic Chondrocytes Cultured in Alginate Beads. **Roum. Biotech. Lett**. (5) 2, 127-134.

108. Mizuno, S. and Glowacki, J. (1996) Three-dimensional composite of demineralized bone powder and collagen for *in vitro* analysis of chondroinduction of human dermal fibroblasts. **Biomaterials** (17) 1819–1825.

109. Allemann, F., Yates, K.E. and Glowacki, J. (2000) Preservation of chondroblastic gene expression signature in 3D culture devices. **J. Bone Miner. Res.** (15) S468.

110. Eid, K., Mizuno, S. and Glowucki, J. (1998) Effect of PTH on matrix production by articular and endochondral chondrocytes in 3D collagen sponges. **Bone** (23) S349.

111. Allemann, F., Mizuno, S., Eid, K., Yates, K.E., Zaleske, D. and Glowacki, J. (2001) Effects of hyaluronan on engineered articular cartilage extracellular matrix gene expression in 3D collagen scaffolds. **J. Biomed. Mater. Res.** (55) 13–19.

112. Herbert, D., Elsworth, R. and Telling, R.C. (1956) The continuous culture of bacteria; a theoretical and experimental study. **J. Gen. Microbiol.** (14) 601-622.

113. Hirtenstein, M. and Clark, J. (1980) Attachment and proliferation of animal cells on microcarriers (CytodexTMl). In *Tissue Culture in Medical Research,* eds. Richards, R.J., Rayan, K.T., Oxford: Pergamon Press, pp 97-104.

114. Posillico, *E.G.* (1986) Microencapsulation technology for large-scale antibody production. **BioTechnology** (4)114-117.

115. Zuhlke, A., Roder, B., Widdecke, H. and Klein, J. (1993) Synthesis and application of new microcarriers for animal cell culture. Part I. Design of, polystyrene based microcarriers. **J. Biomater.** (5) 65-78.

116. Roder, B., Zuhlke, A., Widdecke, H. and Klein, J. (1993) Synthesis and application of new microcarriers for animal cell culture. Part II. Application of polystyrene microcarriers. **J. Biomater.** (5) 79-88.

117. Frondoza, C., Sotiabi, A. and Hungerford, D. (1996) Human chondrocytes proliferate and produce matrix components in microcarrier suspension culture. **Biomaterials** (17) 879-888.

118. Freed, L.E., Vunjak-Novakovic, G. and Langer, R. (1993) Cultivation of cell-polymer cartilage implants in bioreactors. **Cell Biochem.** (51) 257-264.

119. Tang, J.S., Chao, C.F. and Au, M.K. (1994) Growth and metabolism of cultured bone cells using microcarrier and monolayer techniques. **Clin. Orthop. Rel. Res.** (300) 254-258.

120. Hunter, W. (1743) Of the structure and diseases of articulating cartilages. **Philos. Trans.** 470:514–521.

121. Buckwalter, J.A. and Mankin, H.J. (1998) Articular cartilage: tissue design and chondrocyte–matrix interactions. **AAOS Inst Course Lect** (47) 477–486.

122. Coutts, R.D., Sah, R.L. and Amiel, D. (1997) Effect of growth factors on cartilage repair. **AAOS Inst Course Lect** (46) 487–494.

123. Hunziker, E.B. (1997) Articular cartilage repair: are the intrinsic biological constraints undermining this process insuperable? **Osteoarthritis Cartilage** (7) 15–28.

124. Buckwalter, J.A. and Mankin, H.J. (1997) Articular cartilage Part II: degeneration and osteoarthrosis, repair, regeneration, and transplantation. **J Bone Jt. Surg.** (79A) 612–632.

125. Johnson, L.L. (1989) Arthroscopic abrasion arthroplasty historical and pathological perspective: present status. **Arthroscopy** (2) 54-69.

126. Ahsan, T., Lottman, L.M., Harwood, F.L., Amiel, D. and Sah, R.L. (1999) Integrative cartilage repair by ß-aminoproprionitrile. **J Orthop. Res.** (17) 850-857.

127. Ahsan, T. and Sah, R.L. (1999) Biomechanics of integrative cartilage repair. **Osteoarthritis Cartilage** (7) 29-40.

128. Caplan, A.I., Elyaderani, M., Mochizuki, Y., Wakitani, S. and Goldberg, V.M. (1997) Principles of cartilage repair and regeneration. **Clin Orthop** (342) 254-269.

129. Zuger, B.J., Ott, B., Mainil-Varlet. P., Schaffner, T., Clemence. J.F., Weber, H.P. and Frenz, M. (2001) Laser solder welding of articular cartilage: tensile strength and chondrocyte viability. **Lasers Surg Med**. (28) 5, 427-34.

130. O'Driscoll, S.W., Keeley, F.W. and Salter, R.B. (1986) The chondrogenic potential of free autogenous periosteal grafts for biological resurfacing of major full-thickness defects in joint surfaces under the influence of continuous passive motion. An experimental investigation in the rabbit. **J Bone Joint Surg. Am.** Sep (68) 7, 1017-35.

131. Hangody, L., Kish, G., Karpati, Z. Udvarhelyi, I., Szigeti, I. and Bely, M. (1998) Mosaicplasty for the treatment of articular cartilage defects: application in clinical practice. **Orthopedics** (21) 751–756.

132. Goldberg, V.M. and Caplan, A.I. (1999) Biologic restoration of articular surfaces. **AAOS Instruction Course Lectures** (48) 623-627.

133. Green, W.T. (1977) Articular cartilage repair: Behavior of rabbit chondrocytes during tissue culture and subsequent allografting. **Clin. Orhtop. Relat. Res**. (79-A) 612-632.

134. Dozin, B., Malpeli, M., Camardella, L., Cancedda, R. and Pietrangelo, A. (2002) Response of young, aged and osteoarthritic human articular chondrocytes to inflammatory cytokines: molecular and cellular aspects. **Matrix Biol**. Aug; (21) 5, 449-59.

135. Domm, C., Schunke, M., Christesen, K. and Kurz, B. (2002) Redifferentiation of dedifferentiated bovine articular chondrocytes in alginate culture under low oxygen tension. **Osteoarthritis Cartilage** (10) 1, 13-22.

136. Jakob, M., Demarteau, O. and Schafer, D. (2001) Specific growth factors during the expansion and redifferentiation of adult human articular chondrocytes enhance chondrogenesis and cartilaginous tissue formation *in vitro*. **J. Cell Biochem**. (81) 368–377.

137. Wakitani, K., Kimura, T. and Hiruka, A. (1989) Repair of rabbit articular surface with allograft chondrocytes embedded in collagen gel. **J. Bone Joint Surg**. (71) 74-80.

138. Gao, J., Dennis, J.E., Solchaga, L.A., Goldberg, V.M. and Calpan, A.L. (2002) Repair of osteochondral defect with tissue engineered two-phase composite material of injectable calcium phosphate and hyaluranon sponge. **Tissue Engg.** (8) 827-837.

139. Noth, U., Tuli, R., Osyczka, A.M., Danielson, K.G. and Tuan, R.S. (2002) *In vitro* engineered cartilage constructs produced by press-coating biodegradable polymer with human mesenchymal stem cells. **Tissue Engg.** (8) 131-144.

140. Mizuno, S. and Glowacki, J. (1996) Chondroinduction of human dermal fibroblasts by demineralized bone in three-dimensional culture. **Exp Cell Res** (227) 89-97.

141. Bosch, P., Musgrave, D.S. and Lee, J.Y. (2000) Osteoprogenitor cells within skeletal muscle. **J.Orthop. Res.** (18) 933–944.

142. Levy, M.M., Joyner, C.J. and Virdi, A.S. (2001) Osteoprogenitor cells of mature human skeletal muscle tissue: an *in vitro* study. **Bone** (29) 317–322.

143. Zuk, P.A., Zhu, M. and Mizuno, H. (2001) Multilineage cells from human adipose tissue: implications for cell-based therapies. **Tissue Engg**. (7) 211–228.

144. Huang, J.I., Beanes, S.R., Zhu, M., Lorenz, H.P., Hedrick, M.H. and Benhaim, P. (2002) Rat extramedullary adipose tissue as a source of osteochondrogenic progenitor cells. **Plast. Reconstr. Surg.** (109) 1033–1043.

145. Kuznetsov, S.A., Mankani, M.H., Gronthos, S., Satomura, K., Bianco, P. and Robey, P.G. (2001) Circulating skeletal stem cells. **J. Cell Biol**. (153) 1133–1140.

146. Nakayama, N. (2003) Macroscopic cartilage formation with embryonic stem cell derived mesodermal progenitor cells. **J. Cell Sci**. (116) 10, 2015-2028.

147. Buttery, L.D. (2001) Differentiation of osteoblasts and *in vitro* bone formation from murine embryonic stem cells. **Tissue Engg**. (7) 89-99.

148. Tsuchiya, K., Chen, G., Ushida, T., Matstuno, T. and Tateishi, T. (2004) The effect of coculture of chondrocytes with mesenchymal stem cells on their cartilaginous phenotype *in vitro*. **Mater. Sci. Engg C** (24) 391-396.

149. Frenkel, S.R., Toolan, B., Menche. D., Pitman, M. and Pachence, J.M. (1997) Chondrocyte transplantation using a collagen bilayer matrix for cartilage repair. **J. Bone Joint Surg.** (79B) 831–836.

150. Nehrer, S., Breinan, H.A., Ramappa, A., Shortkroff. S., Young, G., Minas, T., Sledge, C.B., Yannas, I.V. and Spector, M. (1997) Canine chondrocytes seeded in type I and type II collagen implants investigated *in vitro*. **J Biomed. Mater. Res.** (38) 95–104.

151. Kang, R., Marui, T., Nita, I., Georgescu, H.I., Suh, J.K., Robbins, R.D. and Evans, C.H. (1997) *Ex vivo* gene transfer to chondrocytes in full-thickness articular cartilage defects-a feasibility study. **Osteoarthritis Cartilage** (5) 139–143.

152. Qi, W.N. and Scully, S.P. (1997) Extracellular collagen modulates the regulation of chondrocytes by transforming growth factor-beta 1. **J Orthop Res** (15) 4, 483–490.

153. Freed, L.E., Grande, D.A., Lingbin, Z., Emmanuel, J., Marquis J.C. and Langer, R. (1994) Joint resurfacing using allograft chondrocytes and synthetic biodegradable polymer scaffolds. **J Biomed. Mater. Res. (**28) 891–899.

154. Agrawal, C.M., Niederauer, G., Micallef, D. and Athanasiou, K. (1995) The use of PLA–PGA polymers in orthopaedics. In: *Encyclopedic Handbook of Biomaterials and Bioengineering*, eds. D. Wise, New York, pp 2081–2115.

155. Freed, L.E., Marquis, J.C., Nohria, A., Emmanual, J., Mikos, A.G. and Langer, R. (1993) Neocartilage formation *in vitro* and *in vivo* using cells cultured on synthetic biodegradable polymers. **J. Biomed. Mater. Res.** (27) 11–23.

156. Chu, C.R., Coutts, R.D., Yoshioka, M., Harwood, F.L., Monosov, A.Z. and Amiel, D. (1995) Articular cartilage repair using allogenic perichondrocyte-seeded biodegradable porous polylactic acid (PLA): a tissue-engineering study. **J. Biomed. Mater. Res.** (29) 1147–1154.

157. Athanasiou, K.A., Niederauer, G.G. and Agrawal, C.M. (1996) Sterilization, toxicity, biocompatibility and clinical applications of polylactic acid/polyglycolic acid copolymers. **Biomaterials** (17) 93–102.

158. Sittinger, M., Reitzel, D., Dauner, M., Hierlemann, H., Hammer, C., Kastenbauer, E., Planck, H., Burmester, G.R. and Bujia, J. (1996) Resorbable polyesters in cartilage engineering: affinity and biocompatibility of polymer fiber structures to chondrocytes. **J.Biomed. Mater. Res.** (33) 57–63.

159. Hanff, G., Sollerman, C., Abrahamsson, S.O. and Lundborg, G. (1990) Repair of osteochondral defects in the rabbit knee with Gore-Tex (expanded polytetrafluoreethylene). An experimental study. **Scand J Plast Reconstr Surg** (24) 217–223.

160. Reisses, N., Downes, S. and Bentley, G. (1994) Cartilage repair using a new synthetic polymer. **Transactions of the Orthopaedics Research Society**, New Orleans, LA.

161. Messner, K. and Gillquist, J. (1993) Synthetic implants for the repair of osteochondral defects of the medial femoral condyle: a biomechanical and histological evaluation in the rabbit knee. **Biomaterials (**14) 513–521.

162. Thomson, R.C., Wake, M.C., Yaszemski, M.J. and Mikos, A.G. (1995) Biodegradable polymer scaffolds to regenerate organs. **Adv. Polymer. Sci.** (122) 245–274.

163. Fuss, M., Ehlers, E., Russlies, M., Rohwedel, J. and Behrens, P. (2000) Characteristics of human chondrocytes, osteoblasts and fibroblasts seeded onto a type I/III collagen sponge under different culture conditions. A light, scanning and transmission electron microscopy study. **Anat. Anz.** (182) 303–310.

164. Ronziere, M. C., Roche, S., Gouttenoire, J., Demarteau, O., Herbage, D. and Freyria, A.(2003) Ascorbate modulation of bovine chondrocyte growth, matrix protein gene expression and synthesis in three-dimensional collagen sponges. **Biomaterials** (24) 851–856.

165. Wakitani, S., Goto, T., Pineda, S., Young, R., Mansour, J., Caplan, A. and Goldberg, V. (1994) Mesenchymal cell-based repair of large, full-thickness defects of articular cartilage. **J. Bone Joint Surg.** (76A) 579–592.

166. Samuel, R. E., Lee, C., Chivizzani, S., Evans, C., Yannas, I., Olsen, B. and Spector, M. (2002) Delivery of plasmid DNA to articular chondrocytes via novel collagen-glycosaminoglycan matrices. **Hum. Gene Ther.** (13) 791–802.

167. Ohno, T, Tanisaka, K, Hiraoka, Y, Ushida, T, Tamaki,T and Tateishi, T. (2004) Effect of type I and type II collagen sponges as 3D scaffolds for hyaline cartilage-like tissue regeneration on phenotypic control of seeded chondrocytes in vitro. **Mat. Sci. Engg.** (24) 3, 407-411.

168. Chen, T., Sato, T., Ushida, R., Hirochika, A. and Tateishi T. (2003) Redifferentiation of dedifferentiated bovine chondrocytes when cultured *in vitro* in a PLGA-type I collagen hybrid mesh. **FEBS Lett**. (542) 95–99.

169. Van Susante, P., Buma, G. and Van Osch. (1995) Culture of chondrocytes in alginate and collagen carrier gels. **Acta Orthop Scand** (66) 6, 549–556.

170. Frondoza, A. Sohrabi, C. and Hungerford, D. (1996) Human chondrocytes proliferate and produce matrix components in microcarrier suspension culture. **Biomaterials** 179, 879–888.

171. Van Damme, M.P., Boyaci, N. and Ilic M. (1998) The size of aggrecan core protein synthesized in monolayer culture and in collagen gels with time in culture. **Int. Cartilage Rep. Soc. Newslett.** (2) 8–9.

172. Lee, C.H., Singla, A.and Lee, Y. (2001) Biomedical application of collagen **Int. J. Pharm**. (221) 1-22

173. Schlag, G. and Redi, H. (1988) Fibrin sealant in orthopaedic surgery. **Clin. Orthop. Rel. Res**. (227) 269-285.

174. Hendrickson, D.A., Nixon, A.J., Grande, D.A., Todhhunter, R.J., Minor, R.M., Erb, H. and Lust, G. (1994) Chondrocytes fibrin matrix transplants for resurfacing extensive articular cartilage defects. **J. Orthop. Res**. (12) 4, 485-497.

175. Ting, V., Sims, C.D., Brecht, L.E., Kasabian, J.G., Connley, A.K., Elisseeff. J, Gittes, G.K. and Longaker, M.T. (1998) *In vitro* prefabrication of human cartilage shapes using fibrin glue and human chondrocytes. **Ann Plast Surg** (40) 4, 413-420.

176. Minas, T. and Peterson, L. (1999) Advanced techniques in autologous chondrocytes transplantation. **Clin. Sports Med.** (18) 1, 13-44

177. Fortier, J.A., Brofman, P.J., Nixon, A.J. and Mohammed, H.O. (1998) Dispartate chondrocyte metabolism in three dimensional fibrin cultures derived from autogenous or commercially manufactured fibrinogens. **Am. J. Vet. Res.** (59) 4 514-520.

178. Kawabe, N. and Yoshinao, M. (1991) The repair of full-thickness articular cartilage defects. Immune responses to reparative tissue formed by allogeneic growth plate chondrocyte implants. **Clin. Orthop**. (268) 279–293.

179. Gelse, K., Von der Mark, K., Aigner, T., Park, J. and Schneider, H. (2003) Articular cartilage repair by gene therapy using growth factor producing mesenchymal cells. **Arthritis Rheum**. (48) 430–441.

180. Piechotta, U. and Flemming, I. (1983) The maximization of wound healing with fibrin glue. **Aesthetic Plast. Surg**. (7) 81–82

181. Rahfoth, B., Weisser, J., Sternkopf, F., Aigner, T., Vonder Mark, K. and Brauer, R. (1998) Transplantation of allograft chondrocytes embedded in agarose gel into cartilage defects of rabbits. **Osteoarthritis Cartilage** (6) 50–65.

182. Diduch, D. R., Jordan, L., Mierisch, C. and Balian, G. (2000) Marrow stromal cells embedded in alginate for repair of osteochondral defects. **Arthroscopy (**16) 571–577

183. Dausse, Y., Grossin, L., Miralles, G., Pelletier, S., Mainard, D., Hubert, P., Baptiste, D., Gillet, P., Dellacherie, E., Netter, P. and Payan. E. (2003) Cartilage repair using new polysaccharidic biomaterials: Macroscopic, histological and biochemical approaches in a rat model of cartilage defect. **Osteoarthritis Cartilage** (11) 16–28.

184. Grigolo, B., Roseti, L. and Fiorini, M. (2001) Transplantation of chondrocyte seeded on a hyaluronan derivative (hyaff-11) into cartilage defects in rabbits. **Biomaterials** (22) 2417-2424.

185. Butnariu- Ephrat, M., Robinson, D., Mendes, D., Halperin, N. and Nevo, Z. (1996) Resurfacing of goat articular cartilage by chondrocytes derived from bone marrow. **Clin. Orthop**. (330) 234-243.

186. Girotto, D., Urbani, S., Brun, P., Renier, D., Barbucci, R. and Abatangelo, G. (2003) Tissue-specific gene expression in chondrocytes grown on three-dimensional hyaluronic acid scaffolds. **Biomaterials** (24) 19, 3265-3275.

187. Chang, C.H., Liu, H.C., Lin, C.C., Chou, C.H. and Lin, F.H. (2003) Gelatin–chondroitin–hyaluronan tri-copolymer scaffold for cartilage tissue engineering. **Biomaterials**, (24) 26, 4853-4858.

188. Sechriest, V.F., Miao, Y.J., Niyibizi, C., Westerhausen-Larson, A., Matthew, H.W., Evans, C.H., Fu, F.H. and Suh, J.K. (2000) GAG-augmented polysaccharide hydrogel: a novel biocompatible and biodegradable material to support chondrogenesis. **J. Biomed. Mater. Res.** (49) 4, 534–541.

189. Solchaga, L.A., Goldberg, V.M. and Caplan, A.I. (2001) Cartilage regeneration using principles of tissue engineering. **Clin. Orthop.** (391S) S161–70.

190. Scully, S.P., Lee, J.W., Ghert, P.M.A. and. Qi, W. (2001) The role of the extracellular matrix in articular chondrocyte regulation. **Clin Orthop** (391S) S72–89.

191. Matthew, H.W., Salley, S.O., Peterson, W.D. and Klein, M.D. (1993) Complex coacervate microcapsules for mammalian cell culture and artificial organ development. **Biotechnol Prog** (9) 510–519

192. Gaserod, O., Smidsrod, O. and Skjak-Braek, G. (1998) Microcapsules of alginate–chitosan-I. A quantitative study of the interaction between alginate and chitosan. **Biomaterials (**19) 1815–1825.

193. Denuziere, A., Ferrier, D., Damour, O. and Domard, A. (1998) Chitosan–chondroitin sulfate and chitosan–hyaluronate polyelectrolyte complexes: biological properties. **Biomaterials (**19) 14, 1275–1285.

194. Lu, F., Prudhommeaux, A., Meunier, L., and Guillemin, G. (1999) Effects of chitosan on rat knee cartilages. **Biomaterials** (20) 20, 1937–1944.

195. Mattioli-Belmonte, M., Gigante, A., Muzzarelli, R.A., Politano, R., De Bendittis, A., Speechia N., Buffa, A., Biagini, G. and Greco, F. (1993) N, N-dicarboxymethyl chitosan as delivery agent for bone morphogenetic protein in the repair of articular cartilage. **Med. Biol. Eng. Comput** (37) 1, 130-134.

196. Chenite, A., Chaput, C., Wang, D., Combes, C., Buschmann, M., Hoemann, C., Leroux, C., Atkinson, B., Binette, F. and Selmani, A. (2000) Novel injectable neutral solutions of chitosan form biodegradable gels *in situ*. **Biomaterials** (21) 2155–2161.

197. Athanasiou, K.A., Schmitz, J.P. and Agrawal, C.M. (1998) The effects of porosity on in vitro degradation of PLA-PGA implants used in repair of articular cartilage. **Tissue Engg.** (4) 53–63.

198. Agrawal, C.M. and Athanasiou, K.A. (1997) A technique to control the pH in the vicinity of biodegrading PLA-PGA implants. **J. Biomed. Mater. Res.** (Appl Biomater) (38) 2, 105–114.

199. Spain, L., Agrawal, C.M. and Athanasiou, K.A. (1998) New technique to extend the useful life of a biodegradable implants. **Tissue Engg.** (4) 343–352.

200. Chu, C. R., Dounchis, J., Yoshioka, M., Sah, R., Coutts, R. D. and Amiel, D. (1997) Osteochondral repair using perichondrial cells. A 1-year study in rabbits. **Clin. Orthop**. (340) 220–229.

201. Dounchis, J. S., Bae, W., Chen, A., Sah, R., Coutts, R. D. and Amiel, D. (2000) Cartilage repair with autogenic perichondrium cell and polylactic acid grafts. **Clin. Orthop**. (377) 248–264.

202. Giurea, A., Klein, T., Chen, A., Goomer, R., Coutts, R. D., Akeson, W., Amiel, D. and Sah, R. (2003) Adhesion of perichondrial cells to a polylactic acid scaffold. **J. Orthop. Res**. (21) 584–589.

203. Caterson, E. J., Li, W., Nesti, L., Albert, T., Danielson, K. and Tuan, R. (2002) Polymer/alginate amalgam for cartilage-tissue engineering. **Ann. N. Y. Acad. Sci**. (961) 134–138.

204. Goomer, R. S., Deftos, L., Terkeltaub, R., Maris, T., Lee, M., Harwood, F. and Amiel, D. (2001) High-efficiency non-viral transfection of primary chondrocytes and perichondrial cells for *ex-vivo* gene therapy to repair articular cartilage defects. **Osteoarthritis Cartilage** (9) 248–256.

205. Frenkel, S. R., Bradica, G., Goldman, S., Kronengold, R., Coutts, R. D. and Brekke, J. H. (2003) Multiphasic device engineered to regenerate articular cartilage. **Trans. Orthop. Res. Soc**. (28) 249.

206. Langer, R. and Vacanti, J. P. (1993) **Tissue Engg. Sci.** (260) 920–926.

207. Cohen, S. B., Meirisch, C., Wilson, H. and Diduch, D. (2003) The use of absorbable co-polymer pads with alginate and cells for articular cartilage repair in rabbits. **Biomaterials** (24) 2653–2660.

208. Sherwood, J. K., Riley, S., Palazzolo, R., Brown, S., Monkhouse, D., Coates, M., Griffith, L., Landeen, L. and Ratcliffe. A. (2002) A three-dimensional osteochondral composite scaffold for articular cartilage repair. **Biomaterials** (23) 4739–4751.

209. Kawanishi, M., Ushida, T., Kaneka, T., Niwa, H., Fukubayashi, T., Nakamura, K., Oda, H., Tanaka, S. and Tateishi, T. (2004) New type of biuodegradable porous scaffolds for tissue engineered articular cartilage. **Mat. Sci. Engg. C** (24) 431-435.

210. Bryant, S.J. and Anseth, K.S. (2002) Hydrogel properties influence ECM production by chondrocytes photo encapsulated in poly (ethylene glycol) hydrogels. **J. Biomed. Mater. Res**. (59) 63-72.

211. Lelah, M.D. and Cooper, S.L. (1986) Polyurethanes in medicine, CRC Press, Boca Raton, FL

212. Gogolewski, S. (1996) Biomedical Polyurethanes. In: *Desk reference of functional polymers. Syntheses and applications,* eds. Arshady, R. (American Chemical Society, Washington) pp. 657–698.

213. Gogolewski, S., Pennings, A.J., Lommen, E., Nieuwenhuis, P. and Wildevuur, Ch.R.H. (1983) Growth of a neo-artery induced by a biodegradable polymeric vascular prosthesis. **Macromol. Chem. Rapid Commun.** (4) pp. 213–219.

214. Gogolewski, S., Walpoth, B. and Rheiner, P. (1987) Polyurethane microporous membranes as pericardial substitutes. **Colloid. Polymer. Sci.** (265) 971–977.

215. Elema, H., de Groot, J.H., Nijenhuis, A.J., Pennings, A.J., Veth, R.P., Klompmaker, J. and. Jansen, H.W. (1990) Use of biodegradable polymer implants in meniscus reconstruction. 2. Biological evaluation of porous biodegradable implants in menisci. **Colloid Polymer Sci.** (268) 1082–1088.

216. Klompmaker, J., Jansen, H.W., Veth, R.P., de Groot, J.H., Nijenhuis, A.J. and Pennings, A.J. (1991) Porous polymer implant for repair of meniscal lesions: a preliminary study in dogs. **Biomaterials** 12 (1991), pp. 810–816.

217. Gogolewski, S., Gorna, K., Rahn, B. and Wieling, R. (2001) Biodegradable polyurethane cancellous bone graft substitute promotes bone regeneration in the iliac crest defects. Transactions 24, 573, **27th Annual Meeting, Society for Biomaterials**, Saint Paul, MN, USA.

218. Gogolewski, S., Gorna, K. and Turner, A.S. (2002) Regeneration of bicortical defects in the iliac crest of estrogen deficient sheep using new biodegradable polyurethane cancellous bone graft substitutes. A pilot study. 48th Annual Meeting, Orthopaedic Research Society, Dallas, TX, USA.

219. Saad, B., Neuenschwander, P., Uhlschmid, G.K. and Suter, U.W. (1999) New versatile, elastomeric, degradable polymeric materials for medicine. **Int. J. Biol. Macromol.** (25) 293–301.

220. Grad, S., Kupcsik, L., Gorna, K., Gogolewski, S. and Alini, M. (2003) The use of biodegradable polyurethane scaffolds for cartilage tissue engineering: potential and limitations. **Biomaterials** (24) 28, 5163-5171.

221. Saeed, K.A., Eribo, B.E., Ayorinde, F.O. and Collier, L. (2002) Characterization of copolymer hydroxybutyrate/hydroxyvalerate from saponified vernonia, soybean, and "spent" frying oils. **J. AOAC. Int.** (85) 4, 917–924.

222. Zinn, M., Witholt, B. and Egli, T. (2001) Occurrence, synthesis and medical application of bacterial polyhydroxyalkanoate. **Adv. Drug Deliver. Rev.** (53), 1, 5–21.

223. Deng, Y, Lin, X.Y., Zheng, Z., Deng, J.G., Chen, J.C., Ma, H. and Chen, G.Q. (2003) Poly (hydroxybutyrate-co-hydroxyhexanoate) promoted production of extracellular matrix of articular cartilage chondrocytes *in vitro.* **Biomaterials** (24) 23, 4273-4281.

224. Lane, J.G., Tontz, W.L. Ball, S.T., Massie, J.B., Chen, A.C., Bae, W.C., Amiel, M.E., Sah, R.L. and Amiel, D. (2001) A Morphologic, biochemical, and biomechanical assessment of short-term effects of osteochondral autograft plug transfer in an animal model. **Arthroscopy** (17) 856–863.

225. Hurtig, M., Evans, P., Clarnette, R., Pearce, S. and Singh, K.M. (2001) Mosaic arthroplasty for resurfacing the equine third carpal bone. **Vet. Surg.** (30) 228–239.

226. Gilbert, J.E. (1998) Current treatment options for the restoration of articular cartilage. **Am J Knee Surg.** (11) 42–46.

227. Pearce, S., Hurtig, M.B., Clarnette, R., Kalra, M. and Miniaci, A. (2001) Investigation of two techniques for optimizing joint surface congruency with mosaic arthroplasty. **J Arthroscopy** (17) 50–55.

228. Pilliar, R.M., Filiaggi, M.J., Wells, J.D., Grynpas, M.D. and Kandel, R.A. (2001) Porous calcium polyphosphate scaffolds for bone substitute applications: *in vitro* characterization. **Biomaterials** (22) 963–972.

229. Waldman, S.D., Grynpas, M.D., Pilliar, R.M. and Kandel, R.A. (2002) Characterization of cartilaginous tissue formed on calcium polyphosphate substrates *in vitro*. **J Biomed. Mater. Res.** (62) 323–330.

230. Bhardwaj, T. Pilliar, R.M., Grynpas, M.D. and Kandel, R.A. (2001) Effect of material geometry on cartilaginous tissue formation *in vitro*. **J Biomed. Mater. Res.** (57) 190–199.

231. Dunn, G., Bellincampi, L.D., Tria, A.J. and Zawadsky, J.P. (1997) Preliminary development of a collagen–PLA composite for ACL reconstruction. **J. Appl. Polym. Sci**. (63) 1423–1428.

232. Chen, G., Ushida, T. and Tatsuya, T. (2000) Hybrid biomaterials for tissue engineering: a preparative method of PLA or PLGA–collagen hybrid sponge. **Adv. Mater**. (12) 455–457.

233. Chen, G., Ushida, T. and Tatsuya, T. (2000) A biodegradable hybrid sponge nested with collagen microsponges. **J. Biomed. Mater. Res**. (51) 273–279.

234. Sato, T., Chen, G., Ushida, T., Ishii, T., Ochiai, N. and Tatsuya, T. (2001) Tissue-engineering cartilage by *in vivo* culturing of chondrocytes in PLGA–collagen hybrid sponge. **Mater. Sci. Engg. C** (17) 83–89.

235. Sellers, R.S., Zhang, R., Glasson, S., Kim, H., Peluso, D., D'Augusta D., Beckwith, K. and Morris, E. (2000) Repair of articular cartilage defects one year after treatment with recombinant human bone morphogenetic protein-2 (rh BMP-2). **J. Bone joint Surg. Am**. (82) 151-160.

236. Marijnissen, W. J., van Osch, G., Aigner, J., Verwoerd-Verhoef, H. and Verhaar, J. (2000) Tissue-engineered cartilage using serially passaged articular chondrocytes. Chondrocytes in alginate, combined *in vivo* with a synthetic (E210) or biologic biodegradable carrier (DBM). **Biomaterials** (21) 571–580.

237. Knudson, W., Casey, B., Nishida, Y., Eger, W., Kuettner, K. and Knudson, C. (2000) Hyaluronan oligosaccharides perturb cartilage matrix homeostasis and induce chondrocytic chondrolysis. **Arthritis Rheumatism** (43) 1165–1174.

238. Grande, D., Pitman, M., Peterson, L., Menche, D. and Klein, M. (1989) The repair of experimentally produced defects in rabbit articular cartilage by autologous chondrocyte transplantation. **J. Orthop. Res**. (7) 208–219.

239. Veldhuijzen, J.P., Huisman, A.H., Vermeiden, J.P. and Prahl-Andersen, B. (1987) The growth of cartilage cells *in vitro* and the effect of intermittent compressive force. A histological evaluation. **Connective Tissue Res.** (16) 187–196.

240. Steinmeyer, S. and Knue, S. (1997) The proteoglycan metabolism of mature bovine articular cartilage explants superimposed to continuously applied cyclic mechanical loading. **Biochem Biophys Res Commun** (240) 216–221.

241. Vunjak-Novakovic, G., Martin, I., Obradovic, B., Treppo, S., Grodzinsky, A.J., Langer R. and Freed L.E. (1999) Bioreactor cultivation conditions modulate the composition and mechanical properties of tissue-engineered cartilage. **J. Orthop. Res.** (17) 130–138.

242. Mauck, R. L., Wang, C., Oswald, E. S. Ateshian, G. A. and Hung, C. T. (2003) The role of cell seeding density and nutrient supply for articular cartilage tissue engineering with deformational loading. **Osteoarthritis and Cartilage** (11) 879–890.

243. LeBaron, R. G. and Athanasiou, K. A. (2000) *Ex vivo* synthesis of articular cartilage. **Biomaterials** (21) 24, 2575-2587.

244. Puelacher, W.C., Kim, S.W., Vacanti, J.P., Schloo, B., Mooney, D. and Vacanti, C.A. (1994) Tissue-engineered growth of cartilage: the effect of varying the concentration of chondrocytes seeded onto synthetic polymer matrices. **Int J. Oral Maxillofacial Surg.** (23) 49–53.

245. Ruoslahti, E. (1984) Fibronectin in cell adhesion and invasion. **Cancer Metastat. Rev.** (3) 43–51.

246. Pierschbacher, M.D. and Ruoslahti, E. (1984) Cell attachment activity of fibronectin can be duplicated by small synthetic fragments of the molecule. **Nature** (309) 30–33.

247. Ruoslahti, E. (1996) RGD and other recognition sequences for integrins. **Ann. Rev. Cell Dev. Biol.** (12) 697–715.

248. Le Baron, R.G. and Athanasiou, K.A. (2000) Extracellular matrix cell adhesion peptides: functional applications in orthopaedic materials. **Tissue Engg** (6) 2, 85–103.

249. Howlett, C.R., Evans, M.D., Walsh, W.R., Johnson, G. and Steel, J.G. (1994) Mechanism of initial attachment of cells derived from human bone to commonly used prosthetic material during cell culture. **Biomaterials** (12) 697-715.

250. Colombet, P., Allard, M., Bousquet, V., de Lavigne, C., Flurin, P.H. and Lachaud, C. (2002) Anterior cruciate ligament reconstruction using four-strand semitendinosus and gracilis tendon grafts and metal interference screw fixation Arthroscopy. **J. Arthroscopic Relat. Surg** (18) 3, 232-237.

251. Yong, W.J.B. and Kwon O.H. (2004) Cross-pin femoral fixation for hamstring posterior cruciate ligament reconstruction Arthroscopy. **J. Arthroscopic Relat. Surg.** (20) 4, 29-33.

252. Hutmacher, D.W. (2000) Scaffolds in tissue engineering bone and cartilage. **Biomaterials** (21) 2529-2543.

253. Bujja, J. (1994) Humoral immune response against minor collagen types IX and X in patients suffering from cartilage graft resorption after reconstructive surgery. **Ann. Rheum. Dis.** (53) 229-234.

254. Rudderman, R., Guyuron, B. and Mendelsohn, B. (1994) The fate of fresh and preserved crushed and non-crushed autogenous cartilage in the rabbit model. **Ann. Plast. Surg.** (32) 250-254.

255. Langer, F. Gross, A.E. and Greaves, M.F. (1972) The auto-immunogenicity of articular cartilage **Exp. Immunol.**(12) 31-7.

256. Rodrigo, J., Sakovich, L., Travis, C. and Smith, G. (1978) Osteocartilaginous allografts as compared with autografts in the treatment of knee joint defects in dogs. **Clin. Orthop. Relat. Res.** (134) 342-349.

257. Garrett, J.C. (1986) Treatment of osteochondral defects of the distal femur with fresh osteochondral allografts: a preliminary report. **Arthroscopy** (2) 222-226.

258. Czitrom, A., Langer, F., McKee, N. and Gross, A. (1986) Bone and cartilage allotransplantation. **Clin. Orthop. Relat. Res.** (8) 141-145.

Advances in Biochemistry and Biotechnology Volume 1 *Pages 53-80*
Edited by: Dr. Chiranjib Chakraborty
Published by: DAYA PUBLISHING HOUSE

2

Human Granulocyte-Colony Stimulating Factor: A Therapeutic Protein

*Shyam Sundar Nandi, Chiranjib Chakraborty, J.K. Deb & Ragini Gothalwal**
*(*Head of the Dept.), School of Biotechnology,*
Guru Ghasidas University, Bilaspur, Koni, (C.G)
E-mail: ragini_gothalwal@hotmail.com

ABSTRACT

Human Granulocyte Colony Stimulating Factor (G-CSF) is a protein of 175 amino acid residues with a molecular weight of 18.8 kda and is encoded by a 525 bp gene encoding. It is one of a family of hematopoietic growth factors, which play an important role in proliferation, and differentiation of cells of the neutrophill lineage. Neutropenia is the most common dose limiting toxicity of conventional chemotherapy. Clinical trails are going on at many places for various newer indications, such as-bone marrow transplantation, treatment in complicated plasmodium falciparum malaria, malignant hematological disease, peripheral blood stem cell transplantation, mylodyplasia syndrome and aplastic anemia, and primary and idiopathic neutropenia. Because of codon bias between prokaryotes and eukaryotes, there becomes a limitation for the expression of hG-CSF in *E.coli* system.rh-GCSF is new proteins pharmaceutical incorporation's version of

* Correspondent Author

hG-CSF. The protein has an amino acid sequence that is identical to the natural G-CSF sequence predicted from human DNA sequence analysis, except for the addition of an N-terminal methionine necessary for expression in *E.coli*. rhG-CSF is produced in *E.coli* by recombinant DNA technology. An attempt to obtain hG-CSF by good cost effective technology in large quantity, by using high expression level vector (pET32a, pET29a, pMALp2, pTrc99A), the gene encoding it has been cloned, and recombinant hG-CSF were similar to those of the natural glycosylated hG-CSF. Therefore, recombinant hG-CSF is commercially produced by a very innovative method to treat patients with neutropenia in a minimum cost. This paper review contains-importance of G-CSF, cytokines, G-CSF, structure of G-CSF, stability and reengineering of G-CSF, kinetics of G-CSF folding, G-CSF and its receptor, atomic structure of the G-CSF-receptor complex, mode of action of G-CSF, gene structure, gene sequence, amino acid sequence, experimental work of rhG-CSF by recombinant DNA technology, G-CSF and its relationship to other growth factor, clinical application of G-CSF, clinical trails going on in G-CSF therapy and the future of this therapeutic protein.

INTRODUCTION

Before we introduce the readers to the epicenter of the chapter *i.e.*, G-CSF we would like to give a brief introduction of "cytokines", the family to which G-CSF belongs. G-CSF happens to be only one of the many cytokines having a vast array of activities in the body and a unbelievably complex and interwoven network of activity. Though we have tried our best to keep the discussion within the boundaries of G-CSF but at instances when a comparison or analogy had to be established other molecules have also been discussed. As we flow down the chapter that we have tried to discuss all the aspects of G-CSF whether factual or applied in the most lucid way possible.

Almost always we say that the term "cytokines" appears in the context of immune response especially "cell-mediated immunity (CMI)" although "cytokines" are likely to be referred to in many different places within the text. Such a practice gained its ground only because the early soluble mediators–the lymphocyte-derived mediators had close association with CMI responses. At a very later stage of research in this direction was the unraveling of a very important fact that many cells of origin other than that of the immune system were also able to produce such factors. Cytokines are a group of low molecular weight regulatory proteins secreted by white blood cells as well as a variety of other cells in the body in response to inducing stimuli and influences the cellular behavior in a big way. But this statement can have an objection now, as we now know that there are membrane-bound forms of some of the secreted cytokines. Cytokines may sometimes be denoted as a hormone like polypeptide, which activates cellular replication, proliferation, and activation in processes relating to host defense and repair. Its role as a regulatory polypeptide which co-ordinates a vast number of muticellular interaction (1).

To be more precise we should mention here that cytokine are polypeptides or glycoproteins having molecular weight of the order of 30 Kda and which are not produced or only feebly produced constitutively. These are mostly local in their action and hence their action can be better called autocrine or paracrine. Cytokine are generally transiently produced and bind to very high affinity receptors on the cells having Ka ranging 10^9 -10^{12} M. Although the range of the activities of an individual cytokine are very diverge and broad but some of their actions are well targeted especially towards hematopoetic cells.

So we can sum up the actions of cytokines by saying that it is a perfect blend of ambiguity by virtue of which it has multiple target cells and multiple actions; Redundancy, as many different cytokines have similar action. Synergism and antagonism are also very common in their action network. A cytokine may also increase or decrease the production and expression of other cytokines and cytokine receptor respectively.

Importance of G-CSF

G-CSF has been my topic of interest not only because of its application as a very potent weapon against diseases such as neutropenia, but its production in bacterial system has always fascinated us a lot. The degree of biological activity obtained in rhG-CSF is very high and generates interest due to the fact that even though there is always some extent of difference in the pattern of glycosylation as it is a protein of eukaryotic origin but their activity *in vivo* is least effected. Not only this, G-CSF has gained a lot of attention of people round the globe because of its cost effectiveness and an relatively affordable therapeutic protein. G-CSF has been successfully produced in the bacterial system and lot of experiments has been performed in which the newly synthesized protein has been targeted to the periplasmic space of the bacteria for the ease of purification. There are also evidences that the protein can be synthesized in the bacteria as inclusion bodies. Thus it may be already clear to the readers of this chapter that we have a preformed propensity and inclination towards the industrial production of G-CSF but we have tried my level best to give as much insight about the molecule into the as possible.

Human Granulocyte Colony Stimulating Factor (2,3) is a protein of 175 amino acid residues with a molecular weight of 18.8 kda encoded by 525 bp gene. Growth and cytokines are potential therapeutic agents of their key roles of regulating the production maturation, activity of the blood; muscles and bone cells.A number of cytokines are currently employed to treatment of blood deficiencies and cancers. For example, erythropoietin is administered to stimulate red blood cell production in patients suffering from anemia. Human growth hormone is used to stimulate growth in children and adults exhibiting extremely slow or subnormal growth due to hormonal deficiencies and granulocyte colony stimulating factor (G-CSF) is employed to treat cancer patients undergoing chemotherapy in order to alleviate the depression of red blood cell produced by cytotoxic therapeutic agents. It is one of a family of hematopoietic growth factors that play an important role in proliferation and differentiation of cells of the neutrophil lineage(4,5,6). Neutropenia is the most common dose limiting toxicity of conventional chemotherapy. Because of codon bias between prokaryotes and eukaryotes, there becomes a limitation for the expression of hG-CSF in *E.coli* system. In this report, the strategy developed involved the use of entire synthetic gene using oligos to increase the expression level. The biological activity of the non-glycosylated recombinant hG-CSF(7) was found to be similar to those of natural glycosylated hG-CSF(8). Neutropenia is the most common dose, limiting toxicity of conventional chemotherapy. Clinical trials are going on at many places for various newer indication, such as-bone marrow transplantation, treatment in complicated *plasmodium falciparum* malaria, malignant hematological disease, peripheral blood stem cell transplantation, mylodyplasia syndrome and aplastic anaemia, primary and idiopathic neutropenia. Recombinant hematopoietic colony-stimulating factors, which stimulate the production of granulocytes, may be a way of reducing the infectious complications of myelotoxic cancer treatment.

Granulocyte colony-stimulating factor (G-CSF) stimulates proliferation of neutrophils and enhances their phagocytic and microcidal activity. Increasing resistance to existing antibacterials and the dearth of new alternatives have complicated the treatment of Gram-negative infections. The aim of this study was to evaluate the efficacy of G-CSF in the treatment of *Pseudomonas aeruginosa* pneumonia when administered in combination with ceftazidime in both neutropenic and non-neutropenic hosts.

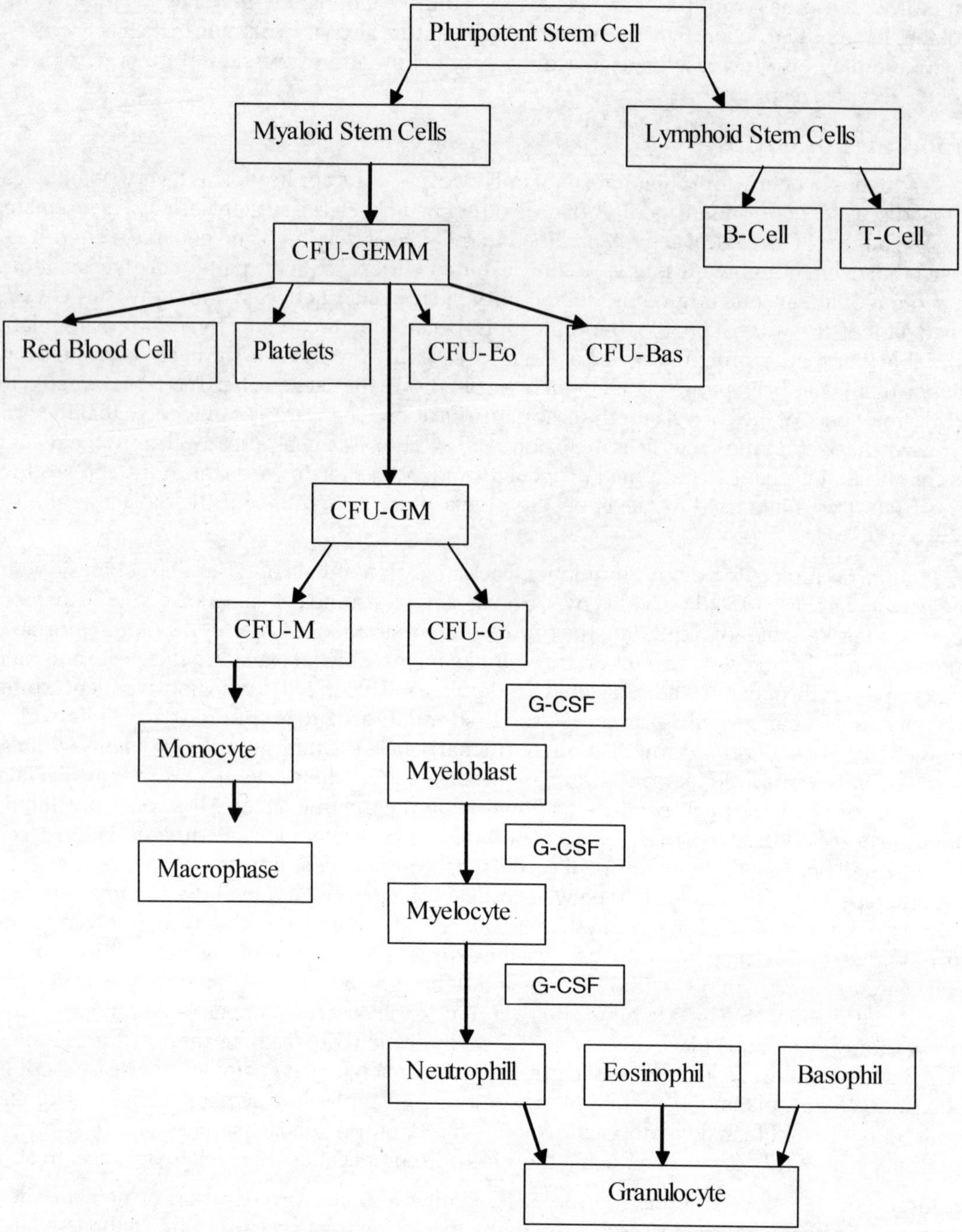

Fig. 2.1: Schematic Illustration of Hematopoiesis. The sites of action of G-CSF in this complex cascade are indicated (11,12)

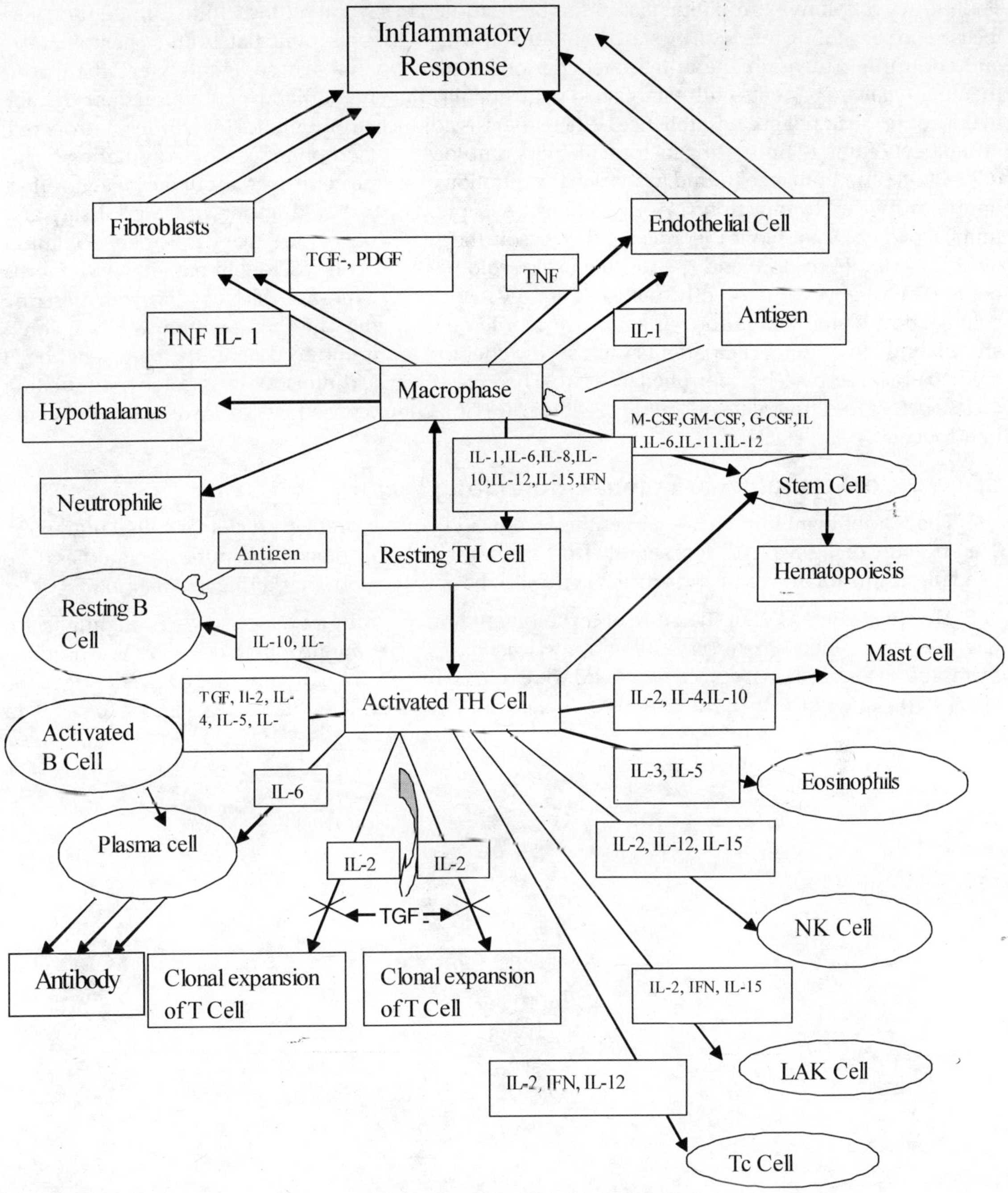

Fig. 2.2: Interaction of Antigen with Nacrophages and the Subsequent Activation of Resting TH Cells Leads to Release of Numerous Cytokines, Generating a Complex Network of Interacting Cells in the Immune Eespons. A small subset of TH cells secretes TGF-β, a cytokine that can inhibit the proliferation of T cells

The efficacy of colony-stimulating factors has been studied in several settings. Initially, attention was focused on preventing infection by starting treatment with the agents immediately after chemotherapy and continuing through the anticipated period of neutropenia. These agents were then used therapeutically in patients with established neutropenia after chemotherapy, although their efficacy in this context had not been established. When given as adjuncts to antibiotics for febrile neutropenia, granulocyte colony-stimulating factor (G-CSF), granulocyte–macrophage colony-stimulating factor (GM-CSF), (9) or both were found to shorten the duration of severe neutropenia, but they had no effect on mortality due to infections or other clinical end points, such as duration of hospital stay and number of days of antibiotic therapy. For this reason, the guidelines of the American Society of Clinical Oncology do not recommend the routine use of colony-stimulating factors in patients with febrile neutropenia. However, a recently published survey revealed that most physicians surveyed, prescribe colony-stimulating factors for patients with febrile neutropenia.The therapeutic use of a colony-stimulating factor in febrile patients with severe chemotherapy-induced neutropenia has not been studied in a large, placebo-controlled trial (10). Theoretically, the administration of colony-stimulating factors before the onset of fever could prevent serious infectious consequences in patients with severe neutropenia.

Structure of Granulocyte Colony Stimulating Factor

The recombinant human G-CSF synthesized in an *E. coli.* expression system is called filgrastim. The structure of filgrastim differs slightly from the natural glycoprotein. Most published studies have used filgrastim and it was the first form of G-CSF to be approved for marketing in Australia.

The three-dimensional structure of recombinant human granulocyte-colony-stimulating factor has been determined by x-ray crystallography (13). Phases were initially obtained at 3.0-Å resolutions by multiple isomorphous replacements and were refined by solvent flattening and by averaging of the electron density of the three molecules in the asymmetric unit. The R factor was 21.5% for all data

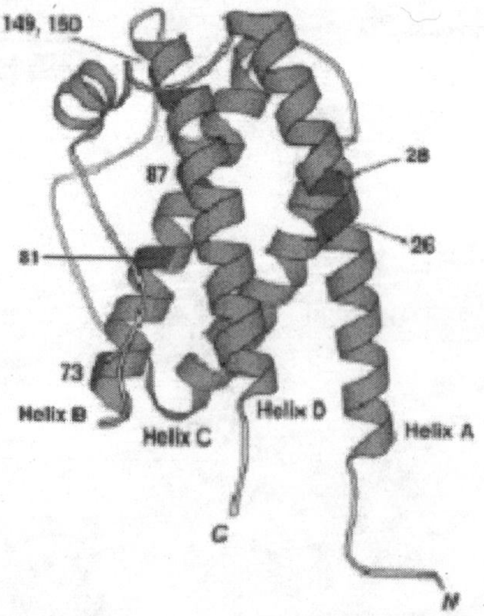

Fig. 2.3: Ribbon Representation of the Structure of G-CSF made in Molscript from NMR Structure (14). A 4-helix bundle motif with long overhand loop connection is shown.The helices are lebeled in order from the N terminus: A,B,C,D.This figure was taken with acknowledgment.

between 6.0- and 2.2-Å resolutions. The structure is predominantly helical, with 104 of the 175 residues forming a four-α-helix bundle. The only other secondary structure is also helical. In the loop between the first two long helices a four-residue 3_{10}-helix is immediately followed by a 6-residue α-helix. Three residues in the short connection between the second and third bundle helices form almost one turn of left-handed helix. The up-up-down-down connectivity with two long crossover connections has been reported previously for five other proteins, which like granulocyte-colony-stimulating factor are all signaling ligands: growth hormone, granulocyte/macrophagecolony-stimulating factor, interferon β, interleukin 2, and interleukin 4. Structural similarity among these growth factors occurs despite the absence of similarity in their amino acid sequences. Conservation of this tertiary structure suggests that these different growth factors might all bind to their respective sequence-related receptors in an equivalent manner.

Stability and Reengineering of Granulocyte Colony-Stimulating Factor

Granulocyte colony-stimulating factor is a long-chain cytokine that has both biological and therapeutic applications. It is involved in the production and maturation of neutrophilic progenitor cells and neutrophils and is administered to stimulate the production of white blood cells to reduce the risk of serious infection in immunocompromised patients. Granulocyte colony-stimulating factor has been reengineered to improve the thermodynamic stability of the protein (13), focusing on enhancing the alpha-helical propensity of residues in the antiparallel 4-helix bundle of the protein.

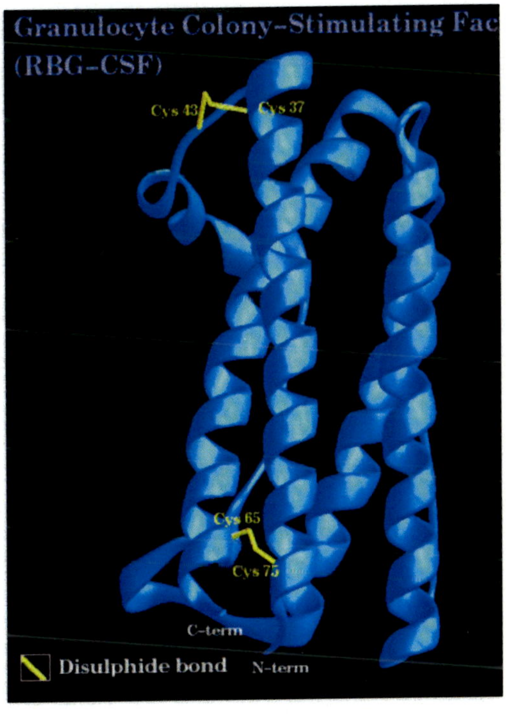

Fig. 2.4
(This figure was taken with acknowledgment)

These redesigns resulted in proteins with substantially enhanced stability while retaining wild-type levels of biological activity, measured as the ability of the reengineered proteins to stimulate the proliferation of murine myeloid cells transfected with the granulocyte colony-stimulating factor receptor

Kinetics of G-CSF Folding

The folding kinetics of G-CSF have been determined by trp-fluorescence and far-UV circular dichroism(15). Folding and unfolding has been achieved by rapid dilution and mixing of the some protein denaturant.G-CSF has a four-helical bundle protein with two long loops between the first and second helices and between the third and fourth helices. The whole conformational change expected by fluorescence was observed by stopped-flow technology, but due to rapid refolding kinetics only a portion was observed by circular dichroism. G-CSF contains two trp residues, and their contribution to the fluorescent-detected kinetics were deciphered through the use of single-site trp mutants. The trp moieties are probes of the local conformation surrounding their environment. One trp at residue 118 is located within the third helix while the other trp at residue 58 is part of the long loop between the first and second helices. The refolding results were most consistent with the following mechanism: $U \leftrightarrow I_1 \leftrightarrow I_2 \leftrightarrow N$; where U represents the unfolded protein, I_1 represents intermediate state 1, I_2 represents intermediate state 2, and N represents the native state. I_1 is characterized as having approximately one-half of the native-like helical structure and none of the native-like fluorescence. I_2 has 100% of the native helical structure and most of the trp-118 and little of the trp-58 native-like fluorescence. Thus refolding occurs in distinct stages with half of the helix forming first followed by the remaining half of the helix including the third helix and finally the loop between the first and second helices folds.

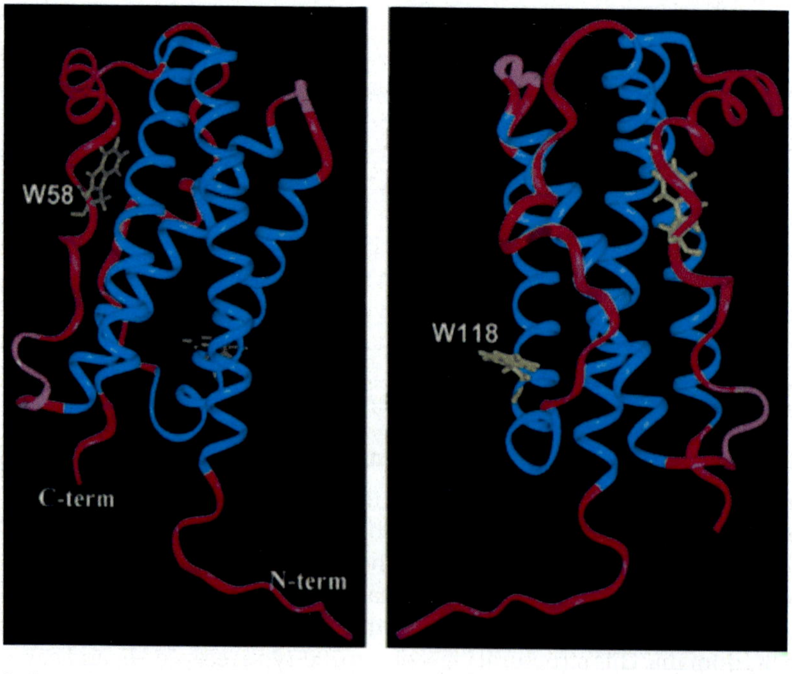

Fig. 2.5: Depiction of the Three-dimensional Structure of the Main Chain of G-CSF as Determined by NMR Spectroscopy (Zink *et al.*, 1994). This figure was taken with acknowledgment.

The four main helices are shown in blue and the loops in purple. The left panel shows the location of W58 within the first long loop, and for the right panel the structure was rotated 180° to illustrate the location of W118 in the third helix.

Granulocyte Colony-stimulating Factor (G-CSF) and its Receptor

Granulocyte colony-stimulating factor (G-CSF) forms a tetrameric complex with its receptor, comprising two G-CSF and two receptor molecules (16). The structure of the complex is unknown, and it is unclear whether there are one or two binding sites on G-CSF and the receptor. The immunoglobulin-like domain and the cytokine receptor homologous module of the receptor are involved in G-CSF binding, and Arg^{288} in the cytokine receptor homologous module is particularly important. To identify residues in G-CSF that interact with Arg^{288}, selected charged residues in G-CSF were mutated to Ala. To clarify whether there are two binding sites, a chimeric receptor was created in which the Ig domain was replaced with that of the related receptor gp130. This chimera bound G-CSF but could not transduce a signal, consistent with failure of dimerization and loss of one binding site. The G-CSF mutants had reduced mitogenic activity on cells expressing wild-type receptor. When tested with the chimeric receptor, all G-CSF mutants except one (E46A) showed reduced binding, suggesting that Glu^{46} is important for interaction with the Ig domain. On cells expressing R288A receptor, all the G CSF mutants except E19A showed reduced mitogenic activity, indicating that Glu^{19} of G-CSF interacts with Arg^{288} of the receptor.

Identification of Ligand-binding of the Granulocyte Colony-stimulating Factor Receptor

The N-terminal Ig-like domain of the G-CSF-R is required for receptor dimerization, but it is not known whether it binds G-CSF or interacts elsewhere in the complex(17). Alanine scanning mutagenesis was used to show that residues in the Ig-like domain of the G-CSF-R (Phe^{75}, Gln^{87}, and Gln^{91}) interact with G-CSF. This binding site for G-CSF overlapped with the binding site of a neutralizing anti-G-CSF-R antibody. A model of the Ig-like domain showed that the binding site is very similar to the viral interleukin-6 binding site (site III) on the Ig-like domain of gp130, a related receptor. To further characterize the G-CSF-R complex, exposed and inaccessible regions of monomeric and dimeric ligand-receptor complexes were mapped with monoclonal antibodies. The results showed that the E helix of G-CSF was inaccessible in the dimeric but exposed in the monomeric complex, suggesting that this region binds to the Ig-like domain of the G-CSF-R. In addition, the N terminus of G-CSF was exposed to antibody binding in both complexes. These data establish that the dimerization interface of the complete receptor complex is different from that in the x-ray structure of a partial complex. A model of the tetrameric G-CSF·G-CSF-R complex was prepared, based on the viral interleukin-6·gp130 complex, which explains these and previously published data.

Atomic Structure of the GCSF-receptor Complex

GCSF is a member of the long-chain subtype of the class 1 cytokine superfamily, which includes growth hormone, erythropoietin, interleukin 6 and oncostatin M. The crystal structure of GCSF complexed to the BN-BC domains, the principal ligand-binding region of the GCSF receptor (GCSFR) has also been determined (18). The two receptor domains form a complex in a 2:2 ratio with the ligand, with a non-crystallographic pseudo-twofold axis through primarily the interdomain region and secondarily the BC domain. This structural view of a gp130-type receptor-ligand complex presents a new molecular basis for cytokine receptor recognition.

The Carboxyl Terminus of the Granulocyte Colony-stimulating Factor Receptor

The G-CSF receptor transduces signals that regulate the proliferation, differentiation, and survival of myeloid cells. A subgroup of patients with severe congenital neutropenia (SCN) has been shown to harbor mutations in the G-CSF receptor gene that resulted in the truncation of the receptor's carboxyl-terminal region. SCN patients with mutations in the G-CSF receptor gene are predisposed to acute myeloid leukemia. The truncated receptors from SCN/acute myeloid leukemia patients mediate augmented and sustained activation of Stat transcription factors and are accordingly hyperactive in inducing cell proliferation and survival but are defective in inducing differentiation. Little is known about the molecular mechanisms underlying the negative role of the receptor's carboxyl terminus in the regulation of Stat activation and cell proliferation/survival. In a study, evidence that SH2-containing phosphatase-1 (SHP-1) plays a negative regulatory role in G-CSF-induced Stat activation was showed. Also, it was demonstrated that the carboxyl terminus of the G-CSF receptor is required for SHP-1 down-regulation of Stat activation induced by G-CSF (11). This regulation is highly specific because SHP-1 has no effect on the activation of Akt and extracellular signal-related kinase1/2 by G-CSF. The data together strongly suggest that SHP-1 may represent an important mechanism by which the carboxyl terminus of the G-CSF receptor down-regulates G-CSF-induced Stat activation and thereby inhibits cell proliferation and survival in response to G-CSF.

Mode of Action of GCSF

Very much like other CSF's, G-CSF initiates its activity by binding to specific all surface receptors which are present in low numbers (1000) on normal neutrophils and promyelocytes. Inflammatory

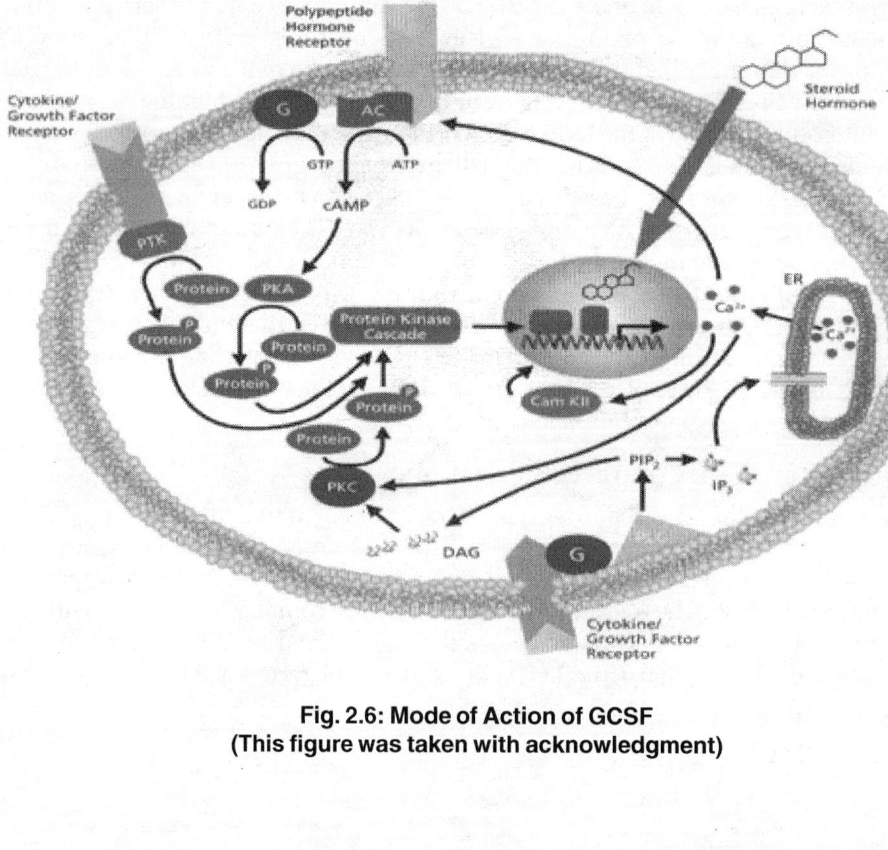

Fig. 2.6: Mode of Action of GCSF
(This figure was taken with acknowledgment)

modulators such as fMet-len-phe,LPS and GM-CSF are the most probable candites which down modulate the G-CSF activity. GCSF receptors of some count and characteristics as that of normal cells are also present on human myeloid leukemia cell line HL60 and have been found consistently on fresh human leukemic cell. most leukemic blasts bearing G-CSF receptors will proliferate in response to the cytokine, but the presence of receptors is not completely predictive of response(19).

G-CSF receptors has only a single high affinity binding site with a dissociation from 60-900 pM is various reports. Cross linking studies suggest that both murine and human G-CSF receptor is a 150 Kda protein and the cytokines, only murine G-CSF can complete to receptor binding of labelled human G-CSF.

G-CSF signal trnsduction mechanism have not yet been fully elucidated. Rapid phophorylation of a 75 Kda protein has been demonatrated after incubation of receptor bearing cells with G-CSF, suggesting tyrosine kinase activity.

Gene Structure

It is reported that the GCSF gene has 4 introns and that 2 different polypeptides are synthesized from the same gene by differential splicing of mRNA.(20) The 2 polypeptides differ by the presence or absence of 3 amino acids. Expression studies indicate that both have authentic GCSF activity.

Some Important Information of GCSF at a Glance

Chromosomal Location

17q11.2-q12.

Micro Array Gene Expression Data

The protein encoded by this gene is a cytokine that controls the production, differentiation, and function of granulocytes. The active protein is found extracellularly. Three transcript variants encoding three different isoforms have been found for this gene. (21)

Swiss Prot Accession No.

P09919 Granulocyte colony-stimulating factor precursor (Homo sapiens); 100% similarity over 207 a.a.

Function

granulocyte/macrophage colony-stimulating factors are cytokines that act in hematopoiesis by controlling the production, differentiation, and function of 2 related white cell populations of the blood, the granulocytes and the monocytes-macrophages. this csf induces granulocytes.

Subcellular Location

Secreted

Subunit

Monomer

Experimental Work of rhGCSF by Recombinant DNA Technology

Granulocyte colony-stimulating factor (G-CSF) is a cytokine that stimulates hematopoiesis and induces proliferation and differentiation of granulocyte progenitor cells as well as production of bone marrow neutrophilic granulocyte colonies. Nowadays, hu-man recombinant G-CSF(hr G-CSF) is used

3'	ACC	CCC	CTG	GGC	CCT	GCC	AGC	TCC	CTG	CCC	CAG	AGC	TTC	CTG
	Thr	Pro	Leu	Gly	Pro	Ala	Ser	Ser	Leu	Pro	Gln	Ser	Phe	Leu
	CTC	AAG	TGC	TTA	GAG	CAA	GTG	AGG	AAC	ATC	CAG	GGC	GAT	GGC
	Leu	Lys	Cys	Leu	Glu	Gln	Val	Arg	Lys	Ile	Gln	Gly	Asp	Gly
	GCA	GCG	CTC	CAG	GAG	AAG	CTG	TGT	GCC	ACC	TAC	AAG	CTG	TGC
	Ala	Ala	Leu	Gln	Glu	Lys	Leu	Cys	Ale	Thr	Tyr	Lys	Leu	Cys
	CAC	CCC	GAG	GAG	CTG	GTG	CTG	CTC	GGA	CAC	TCT	CTG	GGC	ATC
	His	Pro	Glu	Glu	Leu	Val	Leu	Leu	Gly	His	Ser	Aeu	Gly	Ile
	CCC	TGG	GCT	CCC	CTG	AGC	TCC	TGC	CCC	AGC	CAG	GCC	CTG	CAG
	Pro	Trp	Ala	Pro	Leu	Ser	Ser	Cys	Pro	Ser	Gln	Ala	Leu	Gln
	CTG	GCA	GGC	TGC	TTG	AGC	CAA	CTC	CAT	AGC	GGC	CTT	TTC	CTC
	Leu	Ala	Gly	Cys	Leu	Ser	Gln	Leu	His	Ser	Gly	Leu	Phe	Leu
	TAC	CAG	GGG	CTC	CTG	CAG	GCC	CTG	GAA	GGG	ATA	TCC	CCC	GAG
	Tyr	Gln	Gly	Leu	Leu	Gln	Ala	Leu	Glu	Gly	Ile	Ser	Pro	Glu
	TTG	GGT	CCC	ACC	TTG	GAC	ACA	CTG	CAG	CTG	GAC	GTC	GCC	GAC
	Leu	Gly	Pro	Thr	Leu	Asp	Thr	Leu	Gln	Leu	Asp	Val	Ala	Asp
	TTT	GCC	ACC	ACC	ATC	TGG	CAG	CAG	ATG	GAA	GAA	CTG	GGA	ATG
	Phe	Ala	Thr	Thr	Ile	Trp	Gln	Gln	Met	Glu	Glu	Leu	Gly	Met
	GCC	CCT	GCC	CTG	CAG	CCC	ACC	CAG	GGT	GCC	ATG	CCG	GCC	TTC
	Ala	Pro	Ala	Leu	Gln	Pro	Thr	Gln	Gly	Ala	Met	Pro	Ala	Phe
	GCC	TCT	GCT	TTC	CAG	CGC	CGG	GCA	GGA	GGG	GTC	CTG	GTT	TCT
	Ala	Ser	Ala	Phe	Gln	Arg	Arg	Ala	Gly	Gly	Val	Leu	Val	Ala
	AGC	CAT	CTG	CAG	AGC	TTC	CTG	GAG	GTG	TCG	TAC	CGC	GTT	CTA
	Ser	His	Leu	Gln	Ser	Phe	Leu	Glu	Val	Ser	Tyr	Arg	Val	Leu
				CGC	CAC	CTT	GCG	CAG	CCC	TGA	5'			
				Arg	His	Leu	Ala	Gln	Pro	OPA				

Fig. 2.7: Natural Human G-CSF, DNA Sequence and Amino Acid Sequence

for the treatment of chemotherapy- and radiotherapy-induced neutropenia, and also in patients with bone marrow transplantation(22).

A cDNA of human G-CSF (hG-CSF) was synthesized by PCR fromrecombinant cloning vector, with two altered nucleotides for increasing mRNAstability and overexpression, then inserted into a pET expression vector under the control of T7 promoter and cloned in *E. coli* strain BL21 (DE3).

After culture and induction of recombinant *E. coli* with IPTG, we achieved a high level expression of the hG-CSF, where it represented approximately 35% of the total protein as determined by SDS-PAGE and confirmed by western blotting with polyclonal and monoclonal hG-CSF antibodies.

rhG-CSF was produced in a significantly high quantity with a yield of 35% of total protein as determined by SDS-PAGE. Since simple purification steps easily obtain it, it may be cost-effective, even on an industrial scale.

The synthesis and secretion of recombinant human granulocyte colony-stimulating factor (rhG-CSF) are investigated in fed-batch cultures at high cell concentration of recombinant Saccharomyces cerevisiae, and some important characteristics of the secreted rhG-CSF are demonstrated. Transcription of the recombinant gene is regulated by a GAL1-10 upstream activating sequence (UASG), and the rhG-CSF is expressed in a hybrid fusion protein consisting of signal sequence of Kluyveromyces lactis killer toxin and N-terminal 24 amino acids of human interleukin 1beta. The intracellular KEX2 cleavage leads to excretion of mature rhG-CSF into extracellular culture broth, and the cleavage process seems to be highly efficient. In spite of relatively low copy number the plasmid propagation is stably maintained even at nonselective culture conditions. The rhG-CSF synthesis does not depend on galactose level, whereas the production of extracellular rhG-CSF was significantly enhanced by increasing the inducer concentration above a certain level and also by supplementing the nonionic surfactant to the culture medium, which is notably due to the enhanced secretion efficiency. Various immunoblotting analyses demonstrate that none of the rhG-CSF is accumulated in the cell wall fraction and that a significant amount of intracellular rhG-CSF antibody-specific immunoreactive proteins is located in the ER. A core N-glycosylation at fused IL-1beta fragment is likely to play a critical role in directing the high-level secretion of rhG-CSF, and the O-glycosylation of secreted rhG-CSF seems nearly negligible. (23) Also the extracellular rhG-CSF is observed to exist as various multimers, and the nature of molecular interaction is evidently not the covalent disulfide bridges. The CD spectra of purified rhG-CSF and Escherichia coli-derived standard show that the conformations of both are similar and are almost identical to that reported for natural hG-CSF

Expression and Simple Purification of Recombinant Human Granulocyte Colony-Stimulating Factor in *E. coli*

A human granulocyte colony stimulating factor(hG-CSF) gene was synthesized and inserted the synthetic gene encoding the 175 amino acids of hG-CSF including the initiation codon into a trp expression vector for over expression in E.Coli. A strong expression vector was constructed by using pUC18,digested with ClaI and SalI restriction enzymes, and a 540 bp fragment containing the encoding region and a simple purification procedure including in vitro refolding was established. It is reported that the final productivity of hG-CSF was 500~600 mg/l culture, and purified hG-CSF by cation exchange and chromatofocusing chromatography (24).

Production of Human Granulocyte Colony Stimulating Factor Fused to the Pelb Signal Peptide in *Escherichia coli*

It is reported that to induce the secretion of human granulocyte colony stimulating factor (hG-CSF) gene was fused to the gene encoding the pectate lysase B signal peptide (PelB) of *Erwinia caratovora* (25), and attempt to induce the secretion of hG-CSF by *E.coli*.In addition, the pH-stat fed batch culture was adopted to over produce hG-CSF by the high cell density-culture of recombinant *E.coli*. and consequently,did not succeed in directing the secretion of hG-CSF into the periplasmic space of *E.coli*, but obtained inclusion bodies of hG-CSF fused to the PelB signal peptide in the intracellular compartment.hG-CSF oligomers were synthesized, annealed and ligated to form a synthetic hg-CSF gene were disgned to contain BamHI and NcoI sites, respectively.The gene encoding the hG-CSF terminator was placed downstream of the hG-CSF gene. Plasmid pET-22b(+) was used to insert the hG-CSF fragment into the NcoI/BamHI. In case Dh5α was used for the cloning of the hG-CSF gene. The plasmid and host strain used for the expression of the hG-CSF gene were pET-22(+) and E.coli BL21(DE3) respectively, obtained from Novagen.The expression experiments were carried out as described by Studier *et al.* The recombinant cells were grown in LB midium containg ampicillin at a concentration of 100μg/ml at 37° C. 0.25 mM IPTG was added when the cell concentration reached an optical density(600nm) to induce the T7 promoter, and the cell were harvested after a further 7-9 hours induction. The induced protein was readily detectable on 12% SDS-PAGE of the cell lysate after staing of the gel with Coomassie blue.No secretion of hG-CSF and no processing of the signal peptide from the fusion protein was observed. Instead, the fusion protein, pelB-hG-CSF, at a concentration of 1.7g/l was produced in an inclusion body form. Interestingly, the biological activity of the purified fusion protein was found to be identical to that of mature hG-CSF, indicating that the pelB signal peptide attached to the N-terminal of hG-CSF had no effect on its biological activity.

The human granulocyte colony-stimulating factor (hG-CSF) was efficiently secreted at high levels in fed-batch cultures of recombinant Saccharomyces cerevisiae. However, the secreted recombinant hG-CSF (rhG-CSF) was shown to exist as large multimers in the culture broth due to strong hydrophobic interaction. It was hardly monomerized even by urea at high concentration. This multimer has been reported to diminish specific receptor-binding activity of hG-CSF and causes undesirable problems in the downstream process. When the rhG-CSF was secreted to extracellular broth in the presence of a non-ionic surfactant (Tween 80) in the culture media, the multimerization of the secreted rhG-CSF was efficiently prevented in the fed-batch cultures. Also, the monomer fraction and secreted efficiency of rhG-CSF were significantly increased at the higher culture pH (6.5). Without using any denaturing agents, the secreted rhG-CSF monomer was easily purified with high recovery yield and purity via a simple purification process under acidic conditions, consisting of diafiltration, cation exchange, and gel filtration chromatography. A lyophilization process devoid of intermonomer aggregation was also designed using effective stabilizing agents. (26)

G-CSF and Other Colony Stimulating Factor

Granulocyte-Colony Stimulating Factor (G-CSF), Macrophage-Colony Stimulating Factor(M-CSF), Granulocyte-Macrophage-Colony Stimulating Factor(GM-CSF).

Cytokine protein that stimulate growth and reproduction of certain kinds of blood cells in the bone marrow. Also referred to as growth factors. The production of white blood cells is controlled by colony stimulating factors. (27) Cancer chemotherapy and inherited disorders are among the causes of low white blood counts, which lower resistance to infection. Thus, CSFs are being investigated not only as a way to counteract low white blood cell counts, but also as a way to produce specific types of white blood cells. In addition, there is hope that CSFs can stimulate the body to produce additional bone marrow as well as cause some cancer cells to stop dividing. (28)

GCSF and its Relationship to Other Growth Factor

The three dimensional structure of recombinant granulocyte colony stimulating factor has determined by x-ray crystallography (29). Phases were initially obtained at 3.0-A resolution by multiple isomorphous replacements and were refined by solvent flattening and by averaging of the electron density of the three molecules in the asymmetric unit. The current R factor is 21.5% for all data between

6.0- and 2.2-A resolutions. The structure is predominantly helical, with 104 of the 175 residues forming a four-alpha-helix bundle. The only other secondary structure is also helical. In the loop between the first two long helices a four-residue 3(10)-helix is immediately followed by a 6-residue alpha-helix. Three residues in the short connection between the second and third bundle helices form almost one turn of left-handed helix. The up-up-down-down connectivity with two long crossover connections has been reported previously for five other proteins, which like granulocyte-colony-stimulating factor are all signaling ligands: growth hormone, granulocyte/macrophage-colony-stimulating factor, interferon beta, interleukin 2, and interleukin 4. Structural similarity among these growth factors occurs despite the absence of similarity in their amino acid sequences. Conservation of this tertiary structure suggests that these different growth factors might all bind to their respective sequence-related receptors in an equivalent manner.

Mapping

By somatic cell hybridization and in situ chromosomal hybridization, mapped the GCSF gene to 17q11 in the region of the breakpoint in the 15; 17 translocation characteristic of acute promyelocytic leukemia. Further studies indicated that the gene is proximal to the said breakpoint and that it remains on the rearranged chromosome 17. Southern blot analysis using both conventional and pulsed field gel electrophoresis showed no rearranged restriction fragments. By use of a full-length cDNA clone as a hybridization probe in human-mouse somatic cell hybrids and in flow-sorted human chromosomes, mapped the gene for GCSF to 17q21-q22 by in situ hybridization. Also assigned CSF3 to 17q21 by in situ hybridization. CSF3 and CRYB1 are located on the same large restriction fragment (HGM10) (23).

Another form of recombinant human G-CSF called *lenograstim* is synthesised in Chinese hamster ovary (CHO) cells. As this is a mammalian cell expression system, lenograstim is indistinguishable from the 174 amino acid natural human G-CSF. No clinical or therapeutic consequences of the differences between filgrastim and lenograstim have yet been identified, but there are no formal comparative studies. G-CSF should not be confused with granulocyte macrophage colony stimulating factor (GM-CSF), which is a distinctly different haemopoietic growth factor and is also under clinical development.

Granulocyte colony-stimulating factor (or colony stimulating factor-3) specifically stimulates the proliferation and differentiation of the progenitor cells for granulocytes determined the partial amino acid sequence of purified GCSF protein, and by using oligonucleotides as probes, isolated several G-CSF cDNA clones from a human squamous carcinoma cell line cDNA library. Cloning of human GCSF cDNA shows that a single gene codes for a 177- or 180-amino acid mature protein of molecular weight 19,600.

Indication

Patients with neutropenia have significant morbidity due to bacterial infections. These infections require treatment in hospital with intravenous antibiotics. The vulnerability of neutropenic patients to infections dictates that antibiotics are given at the onset of fever, before microbiological confirmation, and continued until the neutropenia resolves to prevent recurrent infection. G-CSF is the first treatment available to Australian patients, which reliably ameliorates the underlying neutropenia.

Clinical Application of G-CSF

Cancer Chemotherapy

G-CSF (filgrastim) is indicated for the prevention of febrile neutropenia in patients receiving myelosuppressive chemotherapy for non-myeloid malignancies. It reduces the duration and severity of post-chemotherapy neutropenia.(30). In 3 randomised placebo controlled studies of G-CSF (filgrastim) given after chemotherapy to patients with small cell lung cancer or lymphoma, the haematological effects of G-CSF (filgrastim) resulted in a 50% reduction in the number of patients experiencing febrile neutropenia. There were associated reductions in the use of intravenous antibiotics, hospitalisation, and the incidence of microbiologically documented infections. These 3 pivotal studies used significantly myelosuppressive chemotherapy regimens: without G-CSF, the incidence of febrile neutropenia for these regimens was 44-77%.

The recommended dose of G-CSF (filgrastim) for the prevention of febrile neutropenia is 5 micrograms/kg/day, commencing the day after chemotherapy and finishing after 10-14 days, when the post-nadir neutrophil level has reached 10x 10^9/L. G-CSF is repeated with each cycle of chemotherapy. Concurrent administration of G-CSF and chemotherapy is not recommended as some combinations may increase myelo suppression. Subcutaneous and intravenous G-CSF has equivalent potency, but subcutaneous injection is generally recommended, as it is suitable for outpatients.

G-CSF (lenograstim) is also approved for use to reduce the incidence of infection associated with established cytotoxic chemotherapy(31). Similar practical considerations apply to its use as for G-CSF (filgrastim).

Some potentially cost-saving variations to the recommended G-CSF regimen are under investigation. They include lower G-CSF doses (in particular, doses based on vial size rather than body mass) and shorter G-CSF courses (in particular, delayed commencement or earlier cessation of G-CSF). Formal assessments of whether or not these other approaches are as effective as the recommended regimen have not been reported.

At present, G-CSF is not recommended for patients with myeloid malignancies because in vitro data suggest that it may stimulate some myeloid malignancies. The safety and efficacy of G-CSF for infection prevention in patients with myeloid leukaemia is currently under investigation. Other research is investigating if the stimulatory effects of G-CSF on myeloid leukaemias will increase the susceptibility of leukaemic cells to cytotoxic drugs.

The patient should be monitored at least twice per week and the full blood count drawn before that day's G-CSF dose. (In the hour immediately following G-CSF administration, there is a transient reduction in numbers of circulating neutrophils.) All febrile neutropenic post-chemotherapy patients must seek urgent medical attention whether or not they have been receiving G-CSF. General practitioners will probably encounter patients who are self-administering G-CSF during outpatient phases of chemotherapy. They should refer febrile patients urgently to hospital.

The greatest benefit of G-CSF is expected with chemotherapy regimens given with intent to achieve cure or prolonged remission and having an incidence of febrile neutropenia above 40% (*e.g.* treatment regimens for patients with non-Hodgkin's lymphoma, relapsed Hodgkin's disease, germ cell tumours and acute lymphoblastic leukaemia). The validity of correlating the degree and duration of neutropenia with the risk of febrile neutropenia or infection has been demonstrated for G-CSF-treated patients in the pivotal studies with G-CSF (filgrastim). Since G-CSF has similar haematological effects for a variety of chemotherapy regimens including some with lower incidences of febrile neutropenic episodes, patients receiving other chemotherapy regimens for different tumour types are also expected to benefit clinically from G-CSF therapy. G-CSF (filgrastim and lenograstim) are approved for use in all non-myeloid tumour types.

The patient should be monitored at least twice per week and the full blood count drawn before that day's G-CSF dose. (In the hour immediately following G-CSF administration, there is a transient reduction in numbers of circulating neutrophils.) All febrile neutropenic post-chemotherapy patients must seek urgent medical attention whether or not they have been receiving G-CSF. General practitioners will probably encounter patients who are self-administering G-CSF during outpatient phases of chemotherapy. They should refer febrile patients urgently to hospital.

The greatest benefit of G-CSF is expected with chemotherapy regimens given with intent to achieve cure or prolonged remission and having an incidence of febrile neutropenia above 40% (*e.g.* treatment regimens for patients with non-Hodgkin's lymphoma, relapsed Hodgkin's disease, germ cell tumors and acute lymphoblastic leukaemia)(32). The validity of correlating the degree and duration of neutropenia with the risk of febrile neutropenia or infection has been demonstrated for G-CSF-treated patients in the pivotal studies with G-CSF (filgrastim). Since G-CSF has similar haematological effects for a variety of chemotherapy regimens including some with lower incidences of febrile neutropenic episodes, patients receiving other chemotherapy regimens for different tumour types are also expected to benefit clinically from G-CSF therapy. G-CSF (filgrastim and lenograstim) is approved for use in all non-myeloid tumour types.

There are many new applications of G-CSF in autologous hematopoietic support under investigation in Australia. G-CSF can be used to increase the numbers of circulating haemopoietic precursor cells which are then harvested by leucapheresis and used either as a supplement to or a substitute for aspirated bone marrow cells in autologous haemopoietic cell transplants. This approach has resulted in significantly accelerated platelet recovery in addition to neutrophil recovery. The role of post-transplant G-CSF after these peripheral blood progenitor cell autologous transplants is being reassessed. Community practitioners should be aware of these new approaches to haemopoietic support after myeloablative chemotherapy as these patients are now often managed as outpatients for a significant proportion of the period of haemopoietic reconstitution. The transplant physician should be contacted if any management questions arise regarding the care of these patients.

Effects of Recombinant Human Granulocyte Colony-Stimulating Factor on Neutropenia in Patients with Congenital Agranulocytosis

Congenital agranulocytosis is a disorder characterized by severe neutropenia and a profound deficiency of identifiable neutrophil progenitors in bone marrow(33). In an attempt to stimulate neutrophil production and thereby reduce the morbidity and mortality associated with this disease, recombinant human granulocyte colony-stimulating factor (rhG-CSF) in doses of 3 to 60 micrograms per kilogram of body weight per day to five patients with congenital agranulocytosis was administered. In all five patients, an increase in the number of neutrophils was noted eight to nine days after the initiation of the effective dosage (the dose at which the neutrophil count reached 1000 cells per microliter or more and the bone marrow showed granulocyte maturation beyond the myelocyte stage). The absolute neutrophil counts rose from less than 100 to between 1300 and 9500 cells per microliter. Marrow aspirates obtained after 14 days at the effective dosage showed maturation to the mature neutrophil stage. The side effects that were observed were medullary pain, splenomegaly, and an elevation of levels of leukocyte alkaline phosphatase. All five patients have had sustained neutrophil counts of 1000 cells per microliter or more for 9 to 13 months while receiving subcutaneous maintenance therapy. Preexisting chronic infections have resolved clinically, and the number of new infectious episodes and the requirement for intravenous antibiotics have decreased. It was concluded that treatment with rhG-CSF can lead to a large increase in the numbers of functional neutrophils in patients with congenital agranulocytosis.

Severe Chronic Neutropenia

The early observations of the efficacy of G-CSF in patients with severe chronic neutropenia and recurrent infections have been confirmed in a multicentre randomised study of patients with congenital, idiopathic and cyclical neutropenia.G-CSF (filgrastim) reduced the incidence and duration of infection-related events and antibiotic use, and the incidence of hospitalisation. G-CSF (filgrastim) is approved for chronic administration to patients with severe chronic neutropenia.

Treatment of Febrile Neutropenia

Not all patients given chemotherapy for cancer receive G-CSF to prevent febrile neutropenia. A randomised placebo-controlled Australian study evaluated the effects of using G-CSF (filgrastim) as an adjunct to empiric intravenous combination antibiotics for post-chemotherapy febrile neutropenia. G-CSF shortened the duration of neutropenia and febrile neutropenia, but did not significantly shorten the duration of fever or hospitalisation, although there were consistent trends towards these benefits. G-CSF halved the relative risk of prolonged hospitalisation. Subset analyses suggested particular benefits for patients with solid tumours and documented infections.

This study supports the hypothesis that there is a real benefit from adding G-CSF to the standard antibiotic management of febrile neutropenia. It seems improbable that this benefit is as significant to patients as the prevention of febrile neutropenia. However, the relative benefit of G-CSF therapy by these two approaches has not been assessed.(34)

Adverse Effect and Safety

G-CSF has been well tolerated by patients. In the randomized studies of patients having chemotherapy for cancer, the only significant adverse effect of G-CSF has been bone pain, which occurred in 15-20% of patients (approximately twice the incidence of placebo-treated patients). This medullary bone pain is effectively managed with simple oral analgesics such as paracetamol with or without codeine. Stronger analgesics are only very rarely required. Thrombocytopenia may be more pronounced in patients who do not have their chemotherapy dose reduced or delayed because of the effect of G-CSF to ameliorate neutropenia (35), which permits intended doses of chemotherapy drugs to be administered on schedule without dose reductions. (Chemotherapy doses are usually decreased if there is a delayed recovery of the neutrophil count.) The thrombocytopenia is probably a consequence of continued delivery of the intended chemotherapy doses rather than an effect of G-CSF itself. G-CSF has not been associated with the flu-like symptoms or fever, which occur with the administration of other cytokines.

The 'low-grade' adverse events related to G-CSF (filgrastim) administration in the randomized study of patients with severe chronic neutropenia were musculoskeletal and bone pain, headache, rash and asymptomatic increase in spleen size. G-CSF (filgrastim) administration to some of these patients for up to 3 years has remained effective and well tolerated. (36)

Infrequent adverse effects reported include vasculitis, Sweet's syndrome (acute febrile neutrophilic dermatosis) and exacerbation of some pre-existing dermatological conditions. Possible anaphylactoid reactions (incidence <0.001%) have been reported, although these have not been confirmed to be allergic reactions. Antibody formation has not been reported, even during G-CSF (filgrastim) courses of up to 9 months.

Cost and Benefit Consideration

G-CSF is an expensive drug. The cost of G-CSF (filgrastim) for the recommended minimum 10-day schedule for prevention of febrile neutropenia after cancer chemotherapy is A$1504 or $2407 per cycle depending on which of the two vial sizes (300 and 480 micrograms/vial) is used. G-CSF (lenograstim) costs $1283.60 for 1O vials (263 micrograms/vial), which is sufficient to treat a 53 kg person for 10 days at the recommended dose. The cost per microgram is $0.50 for G-CSF (filgrastim) and $0.49 for G-CSF (lenograstim). In Australia, a full assessment of the economic impact of G CSF is difficult because of the different sources of health care funding. A significant proportion of the cost of G-CSF may be offset by the savings to the total health care budget through the reduction in admissions for management of febrile neutropenia. These admissions have a median duration of 8 days and are estimated to cost in the vicinity of A$500-700 per day. Such cost offsetting has been documented within both the private and public health care systems in the U.S.A. For the few patients with severe chronic neutropenia and clinically significant infections for whom there is no reliably effective alternative therapy, the benefits of daily G-CSF administration such as an improved quality of life, the significant reduction in morbidity and the reduced cost of managing infections must be considered in cost-benefit analyses. Cost of per vial (300mg) is Rs.-3000.00 in India. Dr.Reddy's, Intus, and some other companies are marketing G-CSF in India.

G-CSF (filgrastim) is available under Section 100 of the Pharmaceutical Benefits Scheme for out patient use during treatment of patients with selected tumour types (non-Hodgkin's lymphoma(35), germ cell tumours, relapsed Hodgkin's disease, acute lymphoblastic leukaemia, Ewing's sarcoma, rhabdomyosarcoma and neuroblastoma), and after autologous marrow transplantation, and in selected patients with severe chronic neutropenia. As of August 1994, G-CSF (lenograstim) also became available as a pharmaceutical benefit for similar uses.

Although G-CSF reduces the morbidity, which accompanies neutropenia, it is not known whether G-CSF affects the mortality of neutropenic sepsis. Febrile neutropenia has a mortality of 10%, but occurs only in a proportion of patients and varies with the degree of myelosuppression of the various chemotherapy regimens. Although febrile neutropenia is a serious complication of chemotherapy, large numbers of patients would be required before a statistically significant effect of G-CSF on mortality could be detected.

G-CSF is currently used for the haematological support of cancer patients receiving chemotherapy, but it is not an anticancer treatment (37). The use of G-CSF allows patients to receive the intended dose of chemotherapy without reductions or an increased dose. However, it is unknown if this will result in a better outcome for the patient.

Balancing the Benefits and Costs of Colony-stimulating Factors: A Current Perspective

Neutropenia is the most common dose-limiting toxicity associated with cancer chemotherapy. Neutropenic complications, including febrile neutropenia, occur in a substantial proportion of patients with chemotherapy-induced neutropenia. Although neutropenia and its complications can be reduced with colony-stimulating factors (CSFs), it is not considered cost-effective to administer them prophylactically to all patients. The current American Society of Clinical Oncology guidelines recommend primary prophylaxis with the CSFs in patients receiving chemotherapy regimens that produce febrile neutropenia in 40% or more of those treated (38). This recommendation was supported by an early economic analysis based on a decision model incorporating hospital cost data to determine the risk threshold for the cost-saving use of the CSFs. Updating this model to include current estimates of hospitalization costs, indirect costs such as productivity losses, and out-of-pocket patient expenses reduced the risk threshold for cost savings with CSF to 18%. Recent efforts have focused on identifying individual patient characteristics that might be used to target prophylactic CSF in patients who are at greatest risk. Should such factors be identified and validated, assessing each patient's individual risk for neutropenic complications may prove to be a better strategy for the cost-effective use of the CSFs than the risk threshold approach. Preliminary results suggest that the targeted application of CSF support based on a risk model may improve the associated cost savings.

Clinical Trails Going on in G-CSF Therapy

New Treatment Options for Managing Chemotherapy-induced Neutropenia

There are so many benefits of using colony-stimulating factors in cancer patients including prevention of neutropenia and decreasing its severity and duration. However, several areas of controversy regarding chemotherapy-induced neutropenia and the use of colony-stimulating factors remain *e.g.* dose intensity, which colony-stimulating factor to use and when, and administration. Colony-stimulating factors have not been found to increase overall survival time in patients undergoing chemotherapy, but they are used to prevent febrile neutropenia, reduce hospitalizations, and improve quality of life. The methods of using colony-stimulating factors are also evolving, with altered administration showing potential as a means of reducing treatment cost. If the rate of febrile neutropenia for a given combination chemotherapy regimen is < 40%, routine prophylactic use of colony-stimulating factors is not considered cost-effective and is not recommended by the American Society of Clinical Oncology. Pegfilgrastim is filgrastim to which polyethylene glycol has been added to extend the drug's half-life and permit less frequent administration. The issue of dose intensity will continue to be evaluated as new cytotoxic agents and combinations are introduced (39). Better models for identifying patients most likely to benefit from colony-stimulating factors are needed.

Current Management of Chemotherapy-induced Neutropenia: The Role of Colony-stimulating Factors

Cytotoxic chemotherapy suppresses the hematopoietic system, impairing host protective mechanisms. One important consequence, chemotherapy-induced neutropenia (CIN), places patients at risk of developing fever and life-threatening infections. These complications have substantial economic impact and may severely affect the quality of life of patients undergoing treatment of cancer (40). Currently, CIN is managed by delaying and reducing chemotherapy treatment with hematopoietic growth factors and with intravenous antibiotic therapy. Reducing chemotherapy may compromise treatment outcomes in potentially curable malignancies, such as early stage breast cancer and non-Hodgkin's lymphoma. Randomized clinical trials have clearly shown that granulocyte colony-stimulating factor (filgrastim) and the longer-acting pegylated granulocyte colony-stimulating factor (pegfilgrastim), when administered 24 hours after chemotherapy before the occurrence of CIN, are effective in reducing the incidence and severity of neutropenia and its complications, including administration of full doses of chemotherapy without treatment delay.

Colony-stimulating Factors Especially G-CSF for the Management of Neutropenia in Cancer Patients

Neutropenia and its subsequent infectious complications represent the most common dose-limiting toxicity of cancer chemotherapy. Febrile neutropenia (FN) occurs with common chemotherapy regimens in 25 to 40% of treatment-naive patients, and its severity depends on the dose intensity of the chemotherapy regimen, the patient's prior history of either radiation therapy or use of cytotoxic treatment, and comorbidities(41). The occurrence of FN often causes subsequent chemotherapy delays or dose reductions. It may also lengthen hospital stay, increase monitoring, diagnostic and treatment costs, and reduce patient quality of life. A decade after their introduction, colony-stimulating factors (CSFs) such as granulocyte colony-stimulating factor (G-CSF) and granulocyte-macrophage colony-stimulating factor (GM-CSF) are now an integral part of the prevention of potentially life-threatening FN; however, only G-CSF has US Food and Drug Administration approval for use in chemotherapy-induced neutropenia. These adjunctive agents accelerate formation of neutrophils from committed progenitors, thereby reducing the duration and severity of neutropenia. Important uses of CSFs in oncology are prevention of FN after chemotherapy, treatment of febrile neutropenic episodes and support following bone marrow transplantation, and collection of CSF-mobilised peripheral blood progenitor cells. G-CSF is used more frequently than GM-CSF for all of these indications because of fewer associated adverse effects. However, they have demonstrated clinical utility in allowing the delivery of planned chemotherapy dose on schedule, an important clinical goal especially in curative tumour settings. The high cost of these agents limits their widespread use. Current American Society of Clinical Oncology guidelines recommend primary prophylaxis, or first cycle use, with CSFs being confined to patients with > or = 40% risk of FN, which may include elderly patients and other high-risk patients. In addition to the risk of FN, primary prophylaxis should also be considered if the patient has risk factors that place them in the Special Circumstances category. (42) These risk factors may include decreased immune function in patients who are already at an increased risk of infection and pre-existing neutropenia due to disease, extensive prior chemotherapy, or previous irradiation to the pelvis or other areas containing large amounts of bone marrow. Future studies are needed to better define the patients most likely to benefit from CSF therapy, both for prophylaxis and as an adjunct to antibiotics for treatment of FN. Other potential uses include combination therapy with stem cell factors and other cytokines to boost progenitor cell development, maintaining dose intensity of salvage therapy in metastatic cancer patients, and application in patients with pneumonia, diabetic foot infections etc, and a variety of other infectious conditions.

Once-per-cycle Pegfilgrastim (Neulasta) for the Management of Chemotherapy-induced Neutropenia

Filgrastim (r-metHuG-CSF) was approved in the United States in 1991 for use in decreasing the incidence of infection, as manifested by febrile neutropenia, in patients with nonmyeloid malignancies treated with myelosuppressive chemotherapy (43). Colony-stimulating factors such as filgrastim are a significant advance in the supportive care of patients with cancer. However, because of its short half-life, filgrastim requires daily dosing by injection to maintain its effects on the bone marrow. Pegfilgrastim is a longer-acting, self-regulating form of filgrastim created by the covalent linkage of a 20-kd polyethylene glycol molecule to the N-terminal of the filgrastim molecule. The molecular characteristics of pegfilgrastim result in a longer terminal half-life, making once-per-chemotherapy-cycle administration possible. Pegfilgrastim has been found to be 11 times more efficient in chemotherapy as per the following observations.The results from two randomized double-blind phase III clinical trials in patients with breast cancer treated with myelosuppressive chemotherapy showed that a single dose of pegfilgrastim provides neutrophil support comparable with that provided by an average of 11 daily injections of filgrastim. Pegfilgrastim has also been shown to be comparable to filgrastim in reducing neutropenic complications in patients treated with chemotherapy for lymphoma. The safety profile of pegfilgrastim is comparable to that of filgrastim in the clinical settings studied to date. The once-per-cycle administration of pegfilgrastim may improve patient quality of life because it is less disruptive to patients and caregivers, and increase adherence because no doses are missed, thus further advancing the management of chemotherapy-induced neutropenia and its consequences. (44)

Effect of Recombinant Human Granulocyte Colony-stimulating Factor (rhG-CSF) on Chemotherapy-induced Neutropenia in Patients with Lung Cancer

An attempt to evaluate rhG-CSF for preventing and reducing the period of chemotherapy-induced neutropenia, phase II studies of the agent, KRN 8601, have also been conducted in patients receiving chemotherapy for lung cancer. In a cooperative study, 53 patients with lung cancer were fully evaluated. The chemotherapy regimen for the patients enrolled in the study was not specified, but an identical regimen with an identical dose and schedule was mandatory for the first cycle in which the patients did not receive the rG-CSF, and for the following cycles in which they received it after completion of chemotherapy for 14 days consecutively (45). Patients were allocated to receive the agents either at a dose of 100, 200 or 400 micrograms/m2 via intravenous drip infusion; or at as ub cutaneous dose of 25, 75, or 125 micrograms. In the study, all the 9 patients with non-small cell lung cancer received a 3-drug combination of vindesine, ifosfamide, and cisplatin(VIP) at an identical dose throughout the cycles. The rG-CSF was administered on the second and the following cycles at a dose of 100 micrograms/m2, subcutaneously, in the same manner as the above. Myelogram and neutrophil functions, *i.e.*, superoxide anion production, chemotactic, and phagocytic activity, were serially determined in these patients. With intravenous dose of 100 micrograms/m2, the rG-CSF considerably elevated the nadir count of neutrophils and significantly reduced the duration of neutropenia. Subcutaneously administered rhG-CSF at 75 micrograms doses did as with intravenous infusion. The optimal dose of the agent in conventional chemotherapy was estimated to be 100 micrograms/m2 when infused intravenously, and 75-125 micrograms subcutaneously. Subcutaneously administered rG-CSF at a dose of 100 micrograms/m2 did not contribute to spare the nadir count of neutrophils, but contributed toward reducing the period of neutropenia induced by the 3-drug combination which was much more myelosuppressive than conventional regimens. Thus, the optimal dose of the agent should be determined according to the dose-intensity of chemotherapy. Peripheral neutrophils obtained after recovery from VIP-induced neutropenia showed a normal activity in superoxide anion production and mobility when combined with rG-CSF, although the activity showed a trend to remain subnormal in the recovered neutrophils without rG-CSF(46). In conclusion, rG-CSF considerably reduces the neutropenia and possibly reduces infections caused by intensive chemotherapy. Hereafter, clinical trials must determine whether rG-CSF improve the therapeutic outcomes of patients receiving chemotherapy in terms of response rate and patient survival.(47)

Colony-stimulating Factors as an Adjunct to Chemotherapy in Small Cell Lung Cancer

Myelo suppression is the major dose-limiting toxicity of chemotherapy in small cell lung cancer (SCLC). The capacity of colony stimulating factors (CSFs) to stimulate granular neutrophil recovery may be of great value to prevent or cure febrile neutropenia and to increase dose-intensity. Primary CSF administration has been shown to reduce the incidence of febrile neutropenia, hospital admission rate, and antibiotic use subsequent to cyclophosphamidedoxorubicin-high dose etoposide (CDE) chemotherapy (48), without improvement of survival or disease control. Primary CSF administration may be recommended when the expected incidence of febrile neutropenia is at least 40%. This benefit has not been established with less myelosuppressive regimens, such as cisplatin-etoposide (PE), which remains an alternative combination of SCLC when standard doses are used. A trial comparing high-dose CDE + CSF with PE would be of considerable interest. There is currently little clinical basis for the use of CSFs to increase chemotherapy dose-intensity, outside clinical trials. Peripheral blood progenitor cells mobilized with CSFs offer interesting prospects. Further studies, with later initiation, shorter duration or lower doses of CSFs, are warranted to improve the cost-effectiveness of CSFs. CSF therapy in addition to antibiotics is normally not justified in febrile neutropenia, except perhaps in selected patients with sepsis syndromes, hypotension or pneumonia. (49)

Evidence-based Use of Colony-stimulating Factors in Elderly Cancer Patients

Neutropenia and its complications represent the major dose-limiting toxicity of cancer chemotherapy, especially in the elderly. Hematopoietic growth factors have been shown to reduce the severity and duration of febrile neutropenia (FN) and to sustain chemotherapy dose intensity. A systematic review was undertaken of studies of the relationship between age and the risk of neutropenia and its complications. The risk of FN associated with standard regimens increases with age and appears to be greatest during the first cycle of chemotherapy (50). FN continues to have a considerable clinical, economic, and quality-of-life impact on affected individuals. The risk of mortality associated with hospitalization with FN also increases with age but is largely associated with the higher rate of comorbidities observed in the elderly population. Despite increasing evidence that elderly patients experience similar benefit from cancer chemotherapy, reductions in dose intensity often compromise response rates and long-term survival. The hematopoietic growth factors reduce the risk of neutropenic events and the need for reduced dose intensity in elderly cancer patients. Primary prophylaxis with colony-stimulating factors (CSFs) reduces the risk of FN and its complications in elderly patients receiving moderately intensive systemic chemotherapy for responsive malignancies. CSFs also appear to reduce cost and improve quality of life in selected elderly patients receiving chemotherapy.

Recombinant Human Granulocyte-macrophage Colony-stimulating Factor After Combined Chemotherapy in High-grade Non-Hodgkin's Lymphoma: A Randomized Pilot Study

High-grade non-Hodgkin's lymphomas (NHL) can potentially be cured with combination chemotherapy, although the optimum schedules still have to be defined. Clinical trials with intensive chemotherapy are predominantly limited by myelosuppression(51). Here, haematopoetic growth factors open up the possibility of reducing chemotherapy-associated toxicities. In a randomised pilot study, the effects of a recombinant human granulocyte-macrophage colony-stimulating factor (rhGM-CSF) following combined chemotherapy with vincristine, doxorubicin, cyclophosphamide, prednisone and etoposide (VACPE) was investigated. A total of 35 patients with high-grade NHLs were randomised to receive either rhGM-CSF or placebo during the first two chemotherapy cycles and rhGM-CSF for all following cycles. rhGM-CSF was administered at a dosage of 5 micrograms/kg for 10 days or until neutrophils were > 1/nl following chemotherapy. The analyses revealed a significant reduction of neutropenia and duration of neutropenia in the rhGM-CSF group. Adverse events were rare and generally mild apart from one anaphylactoid reaction. No effects of rhGM-CSF were observed concerning the platelet nadir or duration of thrombocytopenia. The benefit of rhGM-CSF for response induction and survival via rhGM-CSF-supported dose intensification remains to be determined.(52)

Effect of Granulocyte-macrophage Colony-stimulating Factor on Neutropenia and Related Morbidity Induced by Myelotoxic Chemotherapy

A phase Ib/II clinical study was undertaken to assess the efficacy of recombinant human granulocyte-macrophage colony-stimulating factor (GM-CSF) to attenuate neutropenia and associated morbidity caused by high-dose anticancer chemotherapy administered in the presence or absence of autologous bone marrow support (53). Around 22 patients were treated with various solid tumors and lymphoid neoplasias with a single daily subcutaneous dose of GM-CSF (250 micrograms/m2)

48 h after a second cycle of highly myelotoxic chemotherapy for a period of 10 days and compared intraindividually neutropenia-related clinical and laboratory variables with data obtained from the same patients having previously received a first neutropenia-inducing cycle of identical chemotherapy in the absence of GM-CSF. The clinical trial showed that GM-CSF is active in neutropenic patients by significantly increasing the neutrophil nadir, reducing the time of relevant neutropenia, and reducing the duration of the patient's hospital stay and necessity for parenteral antibiotics. No significant toxicity was encountered with subcutaneous GM-CSF treatment. (54)

Evidence-based Use of Colony-stimulating Factors in Elderly Cancer Patients

Neutropenia and its complications represent the major dose-limiting toxicity of cancer chemotherapy, especially in the elderly. Hematopoietic growth factors have been shown to reduce the severity and duration of febrile neutropenia (FN) and to sustain chemotherapy dose intensity. A systematic review was undertaken of studies of the relationship between age and the risk of neutropenia and its complications. The risk of FN associated with standard regimens increases with age and appears to be greatest during the first cycle of chemotherapy (50). FN continues to have a considerable clinical, economic, and quality-of-life impact on affected individuals. The risk of mortality associated with hospitalization with FN also increases with age but is largely associated with the higher rate of comorbidities observed in the elderly population. Despite increasing evidence that elderly patients experience similar benefit from cancer chemotherapy, reductions in dose intensity often compromise response rates and long-term survival. The hematopoietic growth factors reduce the risk of neutropenic events and the need for reduced dose intensity in elderly cancer patients(55). Primary prophylaxis with colony-stimulating factors (CSFs) reduces the risk of FN and its complications in elderly patients receiving moderately intensive systemic chemotherapy for responsive malignancies. CSFs also appear to reduce cost and improve quality of life in selected elderly patients receiving chemotherapy.

Conclusion

There are preliminary data suggesting that G-CSF will be useful in the following conditions:

1. Neutropenia due to infections (such as occurs in viral infections)
2. Neutropenia due to non-chemotherapeutic myelosuppressive treatments (such as gancyclovir)
3. Drug-induced agranulocytosis.
4. Primary marrow failure (such as aplastic anaemia)

It will be interesting to see if a role for G-CSF can be demonstrated in non-neutropenic states, such as burns or diabetes (where there are adequate numbers of neutrophils but they are functionally defective), or in the management of infections in patients with normal neutrophil counts.

Growth factors and cytokines are potential therapeutic agents of their key roles of regulating the production maturation, activity of the blood; muscles and bone cells.A number of Colony Stimulating Factor including GCSF are currently employed to treatment of blood deficiencies and cancers. Now we are trying for high yield and cost effective production of rh-GCSF by modern innovative method, which will be beneficial for Indian population.

Acknowledgements

I am very much thankful to Prasantada, Mr. Akash Verma, Mr. Snehasis Jana, Mr. Rajib. Mr. Swapan Patra, Miss Usha, Miss Rupali, Miss Nidhi, Mr. Santosh, Mrs. Sipra for their noble encouragement and help to write this article.

REFERENCES

1. http://www.umdnj.edu/pathnweb/genpath/lec 1/Features Philosophy/Features/features.htm

2. Metcalf, D. (1985) The granulocyte-macrophage colony stimulating factors. **Science**, 229, 16-22.

3. Wingfield.P., Benedict.R., Turcatti.G., Allet. B., Mermod. J.J., Delamarter. J., Simona. M.G., and Rose, (1988) K.Characterization of recombinant derived granulocyte colony stimulating factor (G-CSF). **Biochem. J**.(256), 213-218.

4. Soo-Hyung Kang, Kyu-Heum Na, Jang-Hyeon Park, Choong-II Park, Se-Yong Lee and Young-Ik Lee (1995) High level expression and simple purification of recombinant human granulocyte colony stimulating factor in E.coli. **Biotechnology letters**, (17) 687-692.

5. Johnston EM, Crawford J. (1998) Hematopoietic growth factors in the reduction of chemotherapeutic toxicity.**Semin Oncology**.

6. Stanley.E.R., Jubinsky.P.T. (1984) Factors affecting the growth and differentiation of haemopoietic cells in culture. **Clin. Haematol**. (13).

7. Souza, L.M., Boone, T.C., Gabrilove, J., Lai, P.H., Zsebo, K.M., Hurdock, D.C., Chazin, V.R., Bruszewski, J.,Moore, M.A.S., Mertelsmann, R., and Welte, K. (1986)Recombinant human granulocyte colony stimulating factor: effects on normal and leukemic myeloid cells. **Science**, 32, 61-65.

8. Bong Hyun Chung, Mi-Jin Sohn, Su-Wanoh, Ui-Sun Park, Haryoung Poo, Beom Soo Kim, Mi Jin Yu and Young Ik Lee(1998) Overproduction of Human Granulocyte colony stimulating factor fused to the PelB signal peptide in E.coli. **Journal of Fermentation and Bioengineering.** (85) 443-446.

9. Lieschk*e.g.*J, Burgess. A.W. (1992) Granulocyte colony-stimulating factor and granulocyte-macrophage colony-stimulating factor (1 and 2). **N Engl J Med**.(327) 28-35,99-106.

10. Chouaid.C., Bassinet.L, Fuhrman.C., Monnet.I., Housset.B. (1998) Routine use of granulocyte colony-stimulating factor is not cost-effective and does not increase patient comfort in the treatment of small-cell lung cancer: an analysis using a Markov model, **J Clin Oncol.** 16 (8). 2700-7.

11. Aritomi. M., Kunishima. N., Okamoto. T., Kuroki. R., Ota. Y., Morikawa. K. (1999) Atomic structure of the GCSF-receptor complex showing a new cytokine-receptor recognition scheme, **Nature.** (6754).713-7.

12. Babior.B.M, Stossel. T.P. (1994) **Hematology,A pathological Approach** (3rd Ed).279-281.

13. Hill.C.P, Osslund. T.D., Eisenberg.D. (1993) The structure of granulocyte-colony-stimulating factor and its relationship to other growth factors. **Proc Natl Acad Sci U S A**. 90(11). 5167-71

14. Barney.Bishop., Debbie.C.,Koay, Alan. C., Sartorelli., Lynne. Regan. (2001). Reengineering Granulocyte Colony-stimulating Factor for Enhanced Stability, (276) 33465-33470

15. David. N. Brems. (2002) The kinetics of G-CSF folding, **Protein Science.** (11) 2504-2511.

16. Layton.J.E, Shimamoto.G., Osslund.T., Hammacher.A., Smith.D.K., Treutlein.H.R., Boone.T. (1999) Interaction of granulocyte colony-stimulating factor (G-CSF) with its receptor. Evidence that Glu19 of G-CSF interacts with Arg288 of the receptor, **J Biol Chem.** 274(25). 17445-51

17. Layton. J.E., Hall. N.E., Connell.F., Venhorst.J., Treutlein.H.R. (2001)Identification of ligand-binding site III on the immunoglobulin-like domain of the granulocyte colony-stimulating factor receptor, **J Biol Chem**. 276 (39). 36779-87

18. Dong. F., Qiu. Y., Yi. T., Touw. I.P., Larner. A.C. (2001) The carboxyl terminus of the granulocyte colony-stimulating factor receptor, truncated in patients with severe congenital neutropenia/ acute myeloid leukemia, is required for SH2-containing phosphatase-1 suppression of Stat activation, **J Immunol**. (11).6447-52

19. Kenneth. Kaushansky., Christopher. B.Brown., Stephen. Petersdorf., Hematopoetic colony stimulating factor, 381-382.

20. Kanda, N., Fukushige, S., Murotsu, T., Yoshida, M. C., Tsuchiya, M., Asano, S., Kaziro, Y., Nagata, S. (1987) Human gene coding for granulocyte-colony stimulating factor assigned to the q21-q22 region of chromosome. **Somat. Cell Molec. Genet**. (13) 679-684.

21. Nagata, S., Tsuchiya, M., Asano, S., Yamamoto, O., Hirata, Y., Kubota, N., Oheda, M., Nomura, H.,Yamazaki, T. (1986)The chromosomal gene structure and two mRNAs for human granulocyte colony-stimulating factor. **EMBO J**. (5) 575-581.

22. Fallah. M. J,Akbari., B.Saeedinia., A.R. Karimi.M., *et al.* **(**2003) Overexpression of Recombinant Human Granulocyte Colony-Stimulating Factor in E. coli,Vol 28,No.3

23. Nagata, S., Tsuchiya, M., Asano, S., Kaziro, Y.,Yamazaki, T., Yamamoto, O., Hirata, Y., Kubota, N., Oheda, M., Nomura, H., Ono, M. (1986) Molecular cloning and expression of cDNA for the human granulocyte colony-stimulating factor. **Nature** (319) 415-418.

24. Soo-Hyung Kang., Kyu-Heum Na., Jang-Hyeon Park., Choong-II Park., Se-Yong Lee and Young-Ik Lee. (1995) High Level Expression and Simple Purification of Recombinant Human Granulocyte Colony-Stimulating Factor in *E.coli*.**Biotechnology letters**,(17).687-692.

25. Bong. Hyun. Chung., Mi-Jin. Sohn., Su-Wan Oh., *et al.* (1998) Overproduction of Human Granulocyte Colony Stimulating Factor Fused to the Pelb Signal Peptide in *Escherichia coli*,**Journal of Fermentation and Bioenginnering**, (85).443-446.

26. Tweardy, D. J., Cannizzaro, L. A., Palumbo, A. P., Shane, S.; Huebner, K.,Vantuinen, P., Ledbetter, D. H., Finan, J. B., Nowell, P. C., Rovera, G. (1987) Molecular cloning and characterization of a cDNA for human granulocyte colony-stimulating factor (G-CSF) from a glioblastoma multiforme cell line and localization of the G-CSF gene to chromosome band 17q21. **Oncogene Res.** (1) 209-220.

27. Metcalf, D. (1985) The granulocyte-macrophage colony-stimulating factors. **Science** (229)16-22.

28. Lieschke G.J., Burgess.A.W. (1992) Granulocyte colony-stimulating factor and granulocyte-macrophage colony-stimulating factor (1 and 2). **N Engl J Med** (327) 28-35,99-106.

29. Hill.C.P., Osslund.T.D., Eisenberg.D. (1993) The structure of granulocyte-colony-stimulating factor and its relationship to other growth factors, **Proc Natl Acad Sci U S A**. (11).5167-71.

30. Basser. R.L., Underhill.C., Davis.I., Green. M.D., Cebon.J., Zalcberg. J., MacMillan. J., Cohen.B., Marty.J, Fox. R.M., Begley.C. G. (2000) Enhancement of platelet recovery after myelosuppressive chemotherapy by recombinant human megakaryocyte growth and development factor in patients with advanced cancer. **J Clin Oncol.** (15). 2852-61

31. Frampton. J.E., Yarker. Y.E., Goa.K.L., Lenograstim. (1995) A review of its pharmacological properties and therapeutic efficacy in neutropenia and related clinical settings. **Drugs.** 49(5). 767-93.

32. Gemmati. D., Ongaro. A., Scapoli. G.L., Della. Porta. M., Tognazzo. S., Serino. M.L., Di. Bona.E., Rodeghiero. F., Gilli. G., Reverberi. R., Caruso. A., Pasello.M., Pellati. A., De. Mattei. M. (2004).Common gene polymorphisms in the metabolic folate and methylation pathway and the risk of acute lymphoblastic leukemia and non-Hodgkin's lymphoma in adults. **Cancer Epidemiol Biomarkers Prev**. 13(5).787-94.

33. Cottle. T.E., Fier.C.J., Donadieu. J., Kinsey. S.E. (2002) Risk and benefit of treatment of severe chronic neutropenia with granulocyte colony-stimulating factor. **Semin Hematol**.39(2).134-40

34. Trillet-Lenoir. V., Green. J., Manegold.C., Von Pawel. J., Gatzemeier.U., Lebeau.B., *et al.* (1993) Recombinant granulocyte colony stimulating factor reduces the infectious complications of cytotoxic chemotherapy. **Eur J Cancer** (29A) 319-24.

35. Graham. J. (1994) Lieschke Granulocyte colony stimulating factor (G-CSF) (**Aust Prescr**. (17)96-9.

36. Dale. D.C, Bonilla. M.A., Davis. M.W., Nakanishi. A.M., Hammond. W.P., Kurtzberg. J., *et al.* (1993) A randomized controlled phase III trial of recombinant human granulocyte colony-stimulating factor (filgrastim) for treatment of severe chronic neutropenia. **Blood** (81) 2496-502.

37. Metcalf. D., Morstyn. G. (1991) Colony-stimulating factors. **Biologic therapy of cancer.** 417-44.

38. Lyman.G.H. (2003) Balancing the benefits and costs of colony-stimulating factors: a current perspective, **Semin Oncol**. (30). 10-17.

39. Mucenski.J.W., Shogan. J.E. (2003) Maximizing the outcomes in cancer patients receiving chemotherapy through optimal use of colony-stimulating factor, **J Manag Care Pharm**. (2).10-14.

40. Dale.D. (2003) Current management of chemotherapy-induced neutropenia: the role of colony-stimulating factors, **Semin Oncol**. 3-9.

41. Dale. D.C. (2002) Colony-stimulating factors for the management of neutropenia in cancer patients, **Drugs.** (62) 1-15.

42. Sheridan.W.P., Morstyn. G., Wolf. M., Dodds. A., Lusk. J., Maher. D., *et al.* (1989)Granulocyte colony-stimulating factor and neutrophil recovery after high-dose chemotherapy and autologous bone marrow transplantation. **Lancet** (2)891-95.

43. Crawford.J. (2003) Once-per-cycle pegfilgrastim (Neulasta) for the management of chemotherapy-induced neutropenia. **Semin Oncol**. 24-30.

44. Gisselbrecht. C., Prentice. H.G., Bacigalupo. A., Biron. P., Milpied. N., Rubie. H., *et al.* (1994) Placebo-controlled phase III trial of lenograstim in bone-marrow transplantation. **Lancet**. (343) 696-700

45. Faulds.D., Lewis.N.J., Milne.R.J. (1992)Recombinant granulocyte colony-stimulating factor(rG-CSF): pharmacoeconomic considerations in chemothempy-induced neutropenia. **PharmacoEconomics**(1).231-49.

46. Ohnoshi.T., Ueoka. H., Kodani. T., Kamei. H., Kimura. I. (1990) Effect of recombinant human granulocyte colony-stimulating factor (rG-CSF) on chemotherapy-induced neutropenia in patients with lung cancer, **Gan To Kagaku Ryoho**. (17).911-18.

47. Crawford. J., Ozer. H., Stoller. R., Johnson. D., Lyman. G., Tabbara. I., *et al.* (1991) Reduction by granulocyte colony-stimulating factor of fever and neutropenia induced by chemotherapy in patients with small-cell lung cancer. **N Engl J Med**, (325)164-70.

48. Urban.T., Schuller. M.P., Lebeau. B. (1996) Colony-stimulating factors as an adjunct to chemotherapy in small cell lung cancer, **Eur Respir J**. (9),596-602.

49. Sheridan. W.P., Morstyn.G., Wolf. M., Dodds.A., Lusk.J., Maher.D., *et al.* (1989) Granulocyte colony-stimulating factor and neutrophil recovery after high-dose chemotherapy and autologous bone marrow transplantation. **Lancet** (2) 891-5.

50. Lyman.G.H., Kuderer. N., Agboola. O., Balducci. L. (2003) Evidence-based use of colony-stimulating factors in elderly cancer patients, **Cancer Control**. (10).487-99.

51. Bettelheim. P., Valent. P., Andreeff. M., Tafuri. A., Haimi. J., Gorischek. C., Muhm. M., Sillaber. C., Haas. O., L.Vieder., **Recombinant human granulocyte-macrophage colony-stimulating factor after combined chemotherapy in high-grade non-Hodgkin's lymphoma-a randomized pilot study**. (77).700-711.

52. Pettengell. R., Gurney. H., Radford. J.A., Deakin. D.P., James. R.,Wilkinson. P.M., *et al.* (1992) Granulocyte colony-stimulating factor to prevent dose-limiting neutropenia in non-Hodgkin's lymphoma: a randomized controlled trial. **Blood**.(80)1430-36.

53. Guerci. A., Lederlin. P., Reyes. F., Bordessoule. D., Sebban. C., Tilly. H., Kerneis. Y., Biron. P., Gisselbrecht. C., Herbrecht. R., Coiffier. B. (1996) Effect of granulocyte colony-stimulating factor administration in elderly patients with aggressive non-Hodgkin's lymphoma treated with a pirarubicin-combination chemotherapy regimen. Groupe d'Etudes des Lymphomes de l'Adulte, **Ann Oncol**. (9).966-9.

54. Trillet-Lenoir. V., Green. J., Manegold. C., Von Pawel. J., Gatzemeier. U., Lebeau. B., *et al.* (1993) Recombinant granulocyte colony stimulating factor reduces the infectious complications of cytotoxic chemotherapy. **Eur J Cancer**. (29A)319-24.

55. Edgardo.Rivera.,M. HaimErder.,Moshe Fridman.,Debra Frye.,andGabriel N. Hortobagyi. (2003) First-cycle absolute neutrophil count can be used to improve chemotherapy-dose delivery and reduce the risk of febrile neutropenia in patients receiving adjuvant therapy: a validation study. **Breast Cancer Res.** (5). 114–120

FIGURE REFERENCES

Fig. 2.2) http://academic.evergreen.edu/curricular/viruses/

Fig 2.4) http://www.biochem.szote.u-szeged.hu/astrojan/protein1.htm

Fig. 2.5) www.proteinscience.org/content/vol11/issue10/image/large/

Fig. 2.6) http://www.sigmaaldrich.com/Area_of_Interest/Life_Science/Cell_Signaling Scientific_Resources/Pathway_Slides_Charts/Cytokines_Growth_Factors_and_Hormornes.html

Advances in Biochemistry and Biotechnology Volume 1 *Pages 81-94*
Edited by: Dr. Chiranjib Chakraborty
Published by: DAYA PUBLISHING HOUSE

3

Chemical Analysis of Leaves of *Cestrum diurnum* (L.) (Solanales: Solanaceae) and Seasonal Variation of Some Primary and Secondary Biochemicals

*A. Ghosh & G. Chandra**
Mosquito Research Unit, Department of Zoology, Burdwan University, West Bengal, India

ABSTRACT

The present investigation comprises biochemical analysis of the common shrub *Cestrum diurnum* (Solanales: Solanaceae) leaves at three different physiological stages (young, mature and senescent respectively). Change in the chemical status of some primary biochemicals (such

* Correspondent Author

as total carbohydrate, starch, lipid and protein) and secondary biochemical (phenolics) have been determined to estimate their levels in relation to seasonal growth of the plant. Presence of some secondary biochemicals such as steroids and alkaloids has been ascertained through TLC analysis. The mature leaves of *Cestrum diurnum* were extracted with Hexane-Ether (1:1 in v/v) solvent in Soxhlet apparatus for the isolation of essential oil. The methyl esters of the oil were identified through TLC and analysed by capillary GLC (HP 6890 PLUS GLC AND HP3398 A GC). The main constituents were Palmitic acid (27.62%), Stearic acid (4.62%) and Oleic acid (3.06%). The relative concentrations of some major unidentified fatty acids were also calculated (16.51, 8.16 and 7.69%) according to the retention period in capillary column.

Introduction

Medicinal plants are significant source of synthetic and herbal drugs. The components of the plant that support the disease-preventive and health regulatory roles are termed as functional food ingredients or phytochemicals. For at least five thousand years, humankind has relied on natural products as the primary source of medicine. The drugs were used in crude forms like expressed juice, powder decoction or infusion without knowledge in detailed scientific parameters. Now, at the start of a new millennium it is estimated by the World Health Organisation that 80% of the world's inhabitants must rely on traditional medicines for health care; these traditional medicines are primarily plant based. It is estimated that 74% of the 119 currently most important drugs contain active ingredients from plants used in traditional medicine. Exploring traditional herbal medicines in the context of modern science is the need for optimum and proper utilization of traditional plant drugs. Due to the remarkable development of analytical methodology; *e.g.*, Gas chromatography (GC), High performance liquid chromatography (HPLC) and Mass spectroscopy (MS), it has been possible to characterize even minor molecular weight based phytochemicals.

The genus *Cestrum* (Solanaceae), with more than 300 species, is globally distributed. *Cestrum diurnum* (Solanaceae: Solanales) is a single or multi-stemmed shrub that is also known as Day Jassmine. The bark of the plant is nearly smooth and grey whereas the inner bark is yellow-brown and slightly bitter. The twigs are slender, greenish gray in colour and sometimes drooping. The leaves are oblong, shiny green to yellow green in colour and membranous (15 × 6.5 cm and have petioles about 2.5 cm long). The flowers are small, whitish and are grouped in panicles attached at the bases of the last leaves in each branch. The unripe berries are greenish in colour and develop into elliptical purplish–black when ripe, measuring about 6.5 mm in length (1).

There are several applications of *Cestrum diurnum* in folk medicine but the toxicity of the species to humans and livestock has been reported(2). The leaves contain a calcinogenic glycoside called 1,25–dihydroxyoholocalciferol that leads to a Vitamin D toxicity that results in elevated serum calcium and deposition of calcium in soft tissues. In human the administration may results nausea, vomiting and hallucination effects although the antimicrobial effects of the essential oil of *C.diurnum* have also been established(2).

Quality control of the plant raw materials is the most important challenge in bringing any of the traditional medicines or modern phytomedicines to the acceptance of concerned people. The major problem in quality control is the variation in the quality and quantity of active compounds due to the existence of ecotype pharmacological variations, mainly because of soil condition, nutritional status climatic condition, and seasonal variation. Therefore, pharmaecological evaluation is necessary before therapeutic use and it may bring to light possible development of new pharmacological properties other than the main therapeutic actions under certain ecological or physiological conditions. The objective of the present study was to evaluate the seasonal changes of some important primary and secondary biochemicals of the medicinal plant *Cestrum diurnum* (L.) in different physiological stages and the preparation of a detailed chemical characterization by the application of modern scientific procedures.

Materials and Methods

Collection of Leaves

Random collections of the young, (<1 week old), mature (1-6 weeks old) and senescent (7-9 weeks old) *Cestrum diurnum* (L.) leaves growing in different fields situated at the outskirt of Burdwan (87 °54′ N, 23°16′ E) were done thrice a month for twelve months from January to December 2003. Leaves were initially cleaned in the laboratory with distilled water and soaked with paper towels before each biochemical assay.

Preparation of Leaf Extracts

One gram each of the young, mature and senescent leaves, cut into pieces (0.5–1 cm), was separately plunged in boiling ethanol and ground for 10 min. Leaf extracts were centrifuged at 2000 rpm for 20 min, and the supernatant and pellet were separated. To remove leaf pigment, each supernatant was passed through charcoal powder and filtered through Whatman No-41 (Maidstone,UK) filter paper.

Estimation of Primary and Secondary Biochemicals

Primary Biochzemicals

Estimation of Total Soluble Carbohydrate

Sum of soluble carbohydrate was measured by modified Anthrone method provided by Dubios *et al.* (3) with some modifications. To estimate the total soluble carbohydrates, 1 ml of the clear supernatant plant extract was collected and treated with 4 ml of Anthrone reagent. The carbohydrate content was measured by the preparation of standard curve against a reagent blank of D–glucose in the UV–visible spectrophotometer read at 630 nm.

Estimation of Total Protein

Total protein was measured following the method of Lowry *et al.*, .(4).To estimate the sum of soluble proteins, each kind of pellets separated earlier was suspended in a suitable volume of 5% TCA solution at 0-5 °C in an ice bath for 10 min. Aliquotes (1 ml each) were taken in a centrifuge tubes and 1 ml of 10% TCA was added to each such tube and centrifuged at 5000 rpm for 45 min. The supernatants were discarded and the pellets were re-extracted once with absolute ethanol and twice with hot ethanol–ether mixture, every time discarding the supernatant. The pellet was treated with alkaline copper reagent (alkaline sodium carbonate: copper sulphate–sodium potassium tartarate solution (50: 1 in v/v) and the level of total proteins was measured against a reagent blank of bovine serum albumins by the addition of Folin Ciocaltu reagent in the UV-Visible spectrophotometer read at 750 nm.

Estimation of Total lipid

The lipid content of the leaves of *C.diurnum* was estimated by the procedure of Folch *et al.*, .(5). To determine lipids, 1 g each of the three kinds of fresh weed leaves was separately homogenized in 20 ml chloroform: methanol (CHC13: MeOH) (1:2 v/v) mixture for 10 min in a cell homogenizer. The extracts were filtered after vigorous shaking and each residue was mixed with 25 ml $CHCl_3$: MeOH mixture and stirred for 30 min. To remove any chlorophyll, the mixture was filtered through activated charcoal. The combined filtrate was shaken with 0.9% aqueous sodium chloride solution, and non-lipid contaminants were eliminated. The lipids thus extracted were dried in a desiccator and weighed in an electrobalance (± 0.01mg)

Estimation of Starch

The amount of starch was measured by modified Anthrone method with necessary modification. To estimate the starch components, 0.5 g of the sample was homogenized in hot 80% ethanol to remove the soluble sugars. The residue was homogenized and centrifuged after washing with hot 80% ethanol (repeated five times) till the washing did not give colour with Anthrone reagent. The residue was dried in a water bath and 5.0 ml of distilled water and 6.5 ml of 52% perchloric acid was added to it. The extraction procedure was continued using fresh perchloric acid until 100 ml of the supernatant was collected. To 1 ml of the supernatant (0.2 ml supernatant + 0.8 ml distilled water solution) 4 ml of the Anthrone reagent was added and the amount of starch was measured in the UV-Visible spectrophotometer read at 630nm.

Estimation of Secondary Biochemicals

Estimation of Phenolics

To determine the level of phenolics the process of Bray and Thrope (1954)(6) was adopted. The level of phenols (hydrolysable and condensed tannins) in the leaves was determined from each of the leftovers in the carbohydrate estimations. The extract was taken in a test tube, and to that 1 ml Folin-Ciocalteu reagent and 2 ml of 20% sodium-carbonate solution were added. The solution was diluted up to 25 ml with water and phenol levels were estimated following standard method against a known level of Pyrocatechol.

Detection of Alkaloid and Steroid

Five grams of each of these weed leaves were homogenized in 100 ml MeOH: H_2O (4:1) and filtered. The filtrate was evaporated (<40°C) to 1/10th of its original volume in a Rotary evaporator and acidified with 2M H_2SO_4. It was further extracted with chloroform (×3). The chloroform and aqueous extracts were analyzed as follows: the chloroform of the chloroform extract was partly evaporated, and the material was subjected to TLC in Silica gel and developed with $CHCl_3$: MeOH (1:1) mixture. To detect any steroid present in the leaves, the Liebermann–Burchard reagent was sprayed on the plate. Further more, the aqueous acid extract was basified to pH 10 with NH_4OH and extracted twice with $CHCl_3$: MeOH (3:1). The extract was reduced in volume by air-drying, subjected to TLC in silicagel and developed with $CHCl_3$: MeOH (1:1). Dragendoff reagent was sprayed on the plate and any alkaloid present was observed.

Estimation of Ash and Moisture Contents

Known weights of young mature and senescent leaves were placed separately in a hot-air oven at 50 ± 0.5 C for 72 h, and materials that showed constant dry weights were removed from the oven and weighed. The difference between fresh and dry weights gave moisture contents of each kind of leaves. A portion (140-150 mg/replicate) of each kind of oven-dried leaves was separately ashed in a muffle furnace at 450± 10°C for 20 min. The ash thus produced in each crucible was soaked with 2 ml distilled water and placed in a watch glass for drying at 110°C for 2 h. The respective ash samples were weighed. The percentages of ash and ash-free weight of each sample were determined.

Isolation of Essential Oil

Preparation of Hexane Extract

20 g of freshly colleted mature leaves of *Cestrum diurnum* were plunged in 1 litre of Hexane in Soxhlet apparatus. The temperature during hot extraction was maintained at 40°– 50°C and continued for 72 hours. The leaf extract thus produced was collected and filtered through Whatman No.41 filter paper. The filtrate was evaporated and the essential oil was kept in a desiccator for further study.

Preparation of Methyl Ester for GLC Study

The dried sample was treated with chlorform: perchloric acid mixture (96: 5 v/v) and the solution was placed on a water bath at 50-60°C for 10 min. After cooling, the solution was extracted with PET ether (50 ml at a time) in a separating funnel (repeated thrice). The PET fractions were collected dried over anhydrous Na2SO4 and kept in a rotary evaporator for drying. The concentrated fraction was developed in TLC and methyl esters were identified in Iodine Chamber (Mobile phase–n hexane: ethyl acetate 1: 1 in v/v). The spots were scrapped, dissolved in dehydrated ethanol and the ethanolic solution was boiled on water bath at 55°–60°C for 25 minutes. The ethanol fraction was kept in a 250 ml round bottom flask discarding the silica gel fraction. Above methods were used for the preparation of methyl esters of standard fatty acids such as–Myriastic acid, Palmitic acid, Stearic acid, Oleic acid and Lenoleic acid. The Methyl esters thus prepared were subjected to GCMS study.

Estimation of Fatty Acids

1 gm of the essential oil was taken in a 250 ml round bottom flask and 25 ml of 0.5 M KOH was added in dehydrated ethanol and the flask was kept on a water bath at 55°– 60° C for 20 min. After completion of the saponification procedure, 12.5 ml of concentrated HCL was added and the whole solution was kept in a separating funnel. 50 ml of petroleum ether was added in the separating funnel and after vigorous shaking for 15 min, PET ether fraction was collected in a separate conical flask discarding the aqueous fraction (repeated thrice). The PET fractions were combined and 50 g of anhydrous Na2SO4 was added to it. The whole solution was filtrate and kept in a desiccator. A fraction of the filtrate was dissolved in hexane and undergone TLC on Silica gel G(Sigma,St.Louis.MO,usa) layers (thickness 0.5 cm) which had been prepared using a Unoplan(Shandon,London) coating apparatus with mobile phase hexane: ethyl acetate 1:1, the plate was kept in Iodine chamber and the spot of fatty acid was confirmed by diphenyl pieryl hydrazyl spray.

Analytical Gas Chromatographic Study

The purified fatty acid fraction was analysed directly by GC on a Hewlett Packard (HP;Palo Alto,(A;USA) MODEL HP-6890 PLUS GLC AND HP 3398 A GC CHEMSTATION instrument filled with a column HP-5(Capillary Column-0.32 mm in diameter, length 30 metre).The oven temperature was initially 150°(4 min) and ended with 250° (5 min)

Identification of Components

The identification of each compound was carried out by comparison of RRT (Relative retention time) and mass spectral data obtained with standard methyl esters preparation of Palmitic acid, Stearic acid, Oleic acid, Linoleic acid, Linolenic acid and Myriastic acid (Obtained from Sigma)

Results

Primary Biochemicals

The soluble sum of carbohydrates in the mature leaves was higher than those in the young and senescent leaves during March–August and November–January (Fig. 3.1). The concentrations of soluble carbohydrates in young and senescent leaf were maximum in the month of February and in mature leaves the concentration was highest in the month of March. The concentration of soluble sugar in young and senescent leaves was minimum in the month of May whereas the lowest concentration was found in the month of August for mature leaves.

The concentration of starch in the mature leaves was always higher than those in the young and senescent leaves (Fig. 3.2). The highest concentration was found during the month of May whereas the lowest concentration was found during the month of August in all kind of leaves.

The seasonal distribution of protein revealed that the concentration was always higher in young and mature leaves than that in the senescent leaves (Fig. 3.3). The highest concentration of protein was found in the month of January, February and September for young, mature and senescent leaves respectively. The lowest concentration was found in the month of May in young leaves, in the month of July in mature leaves and in the month of January in senescent leaves.

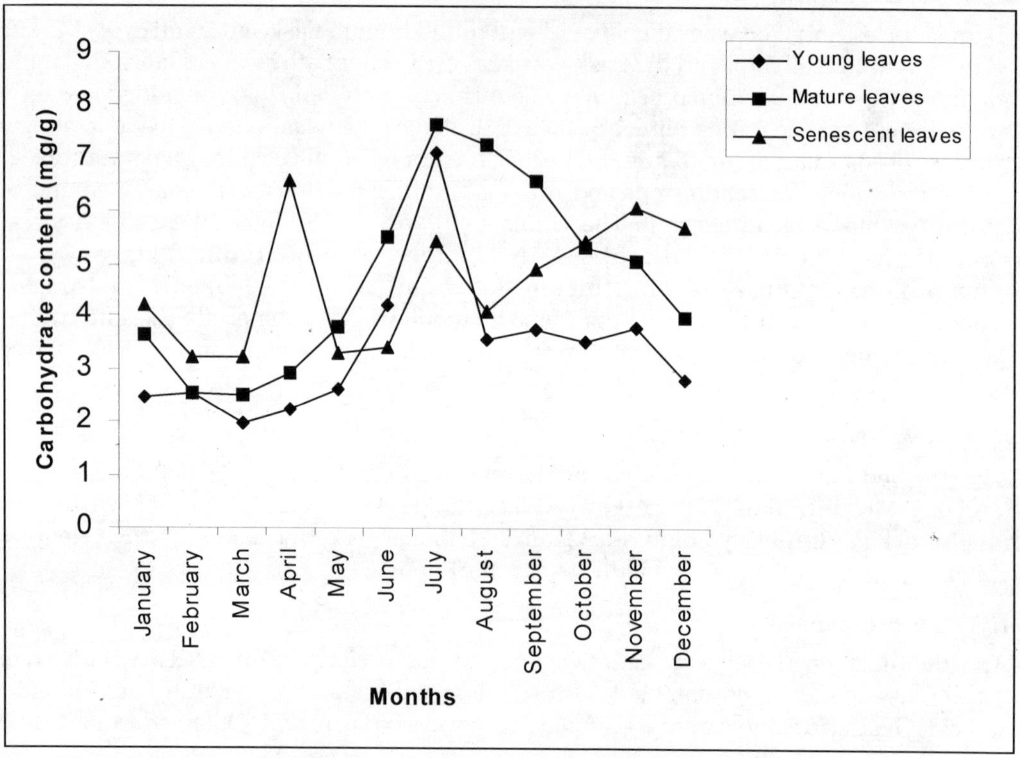

Fig. 3.1: Seasonal Fluctuation in the Level (mg/g) of Carbohydrate Content in the Leaves of *Cestrum diurnum* in Different Physiological Stages

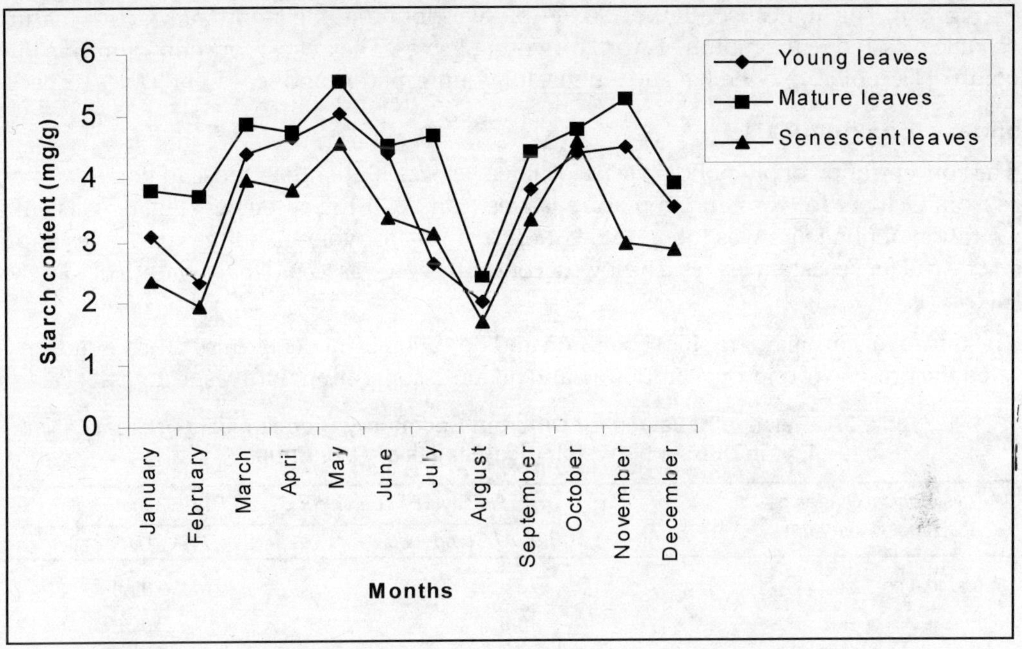

Fig. 3.2: Seasonal Fluctuation in the Level (mg/g) of Starch Content in the Leaves of *Cestrum diurnum* in Different Physiological Stages

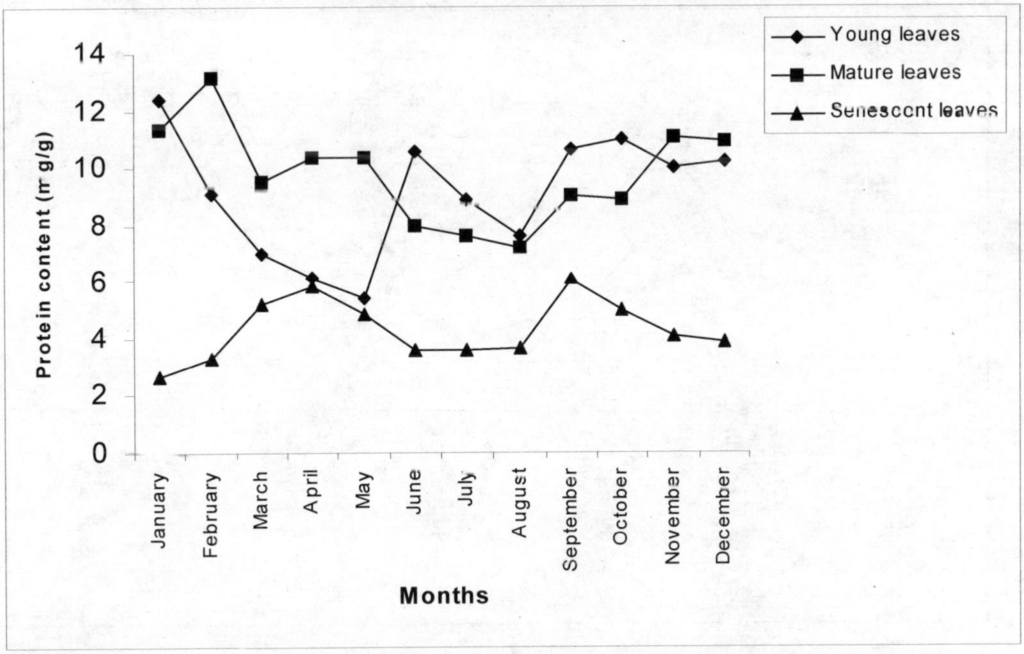

Fig. 3.3: Seasonal Fluctuation in the Level (mg/g) of Protein Content in the Leaves of *Cestrum diurnum* in Different Physiological Stages

The concentration of lipid in mature leaves was always higher than those in young and senescent leaves (Fig. 3.4). The highest concentration of lipid was found in the month of May for mature and senescent leaves and in the month of April for young leaves. The lowest concentration was found in the month of December, November and August for young, mature and senescent leaves respectively.

Secondary Biochemicals

The concentration of phenolics in senescent leaves was higher than those in young and mature leaves from October to April and in mature leaves from May to September (Fig. 3.5). The highest concentration of phenolics was found in the month of July for young and mature leaves and in the month of April for senescent leaves. The lowest concentration was found in the month of March for all kind of leaves.

The thin layer chromatographic separation and detection of biochemicals in three kinds of leaves indicated the presence of steroid compound and alkaloid compound with respective Rf (Table 3.1).

Table 3.1: Relative Rf Values for Different Secondary Biochemicals Present in Different Physiological Stages of *C. diurnum*

Physiological Stages of the Plant C. diurnum	*Relative Rf Length for Secondary Biochemicals*	
	Steroid Compound	*Alkaloid Compound*
Young leaves	0.41	0.58
Mature leaves	0.53	0.73
Senescent leaves	0.67	0.48

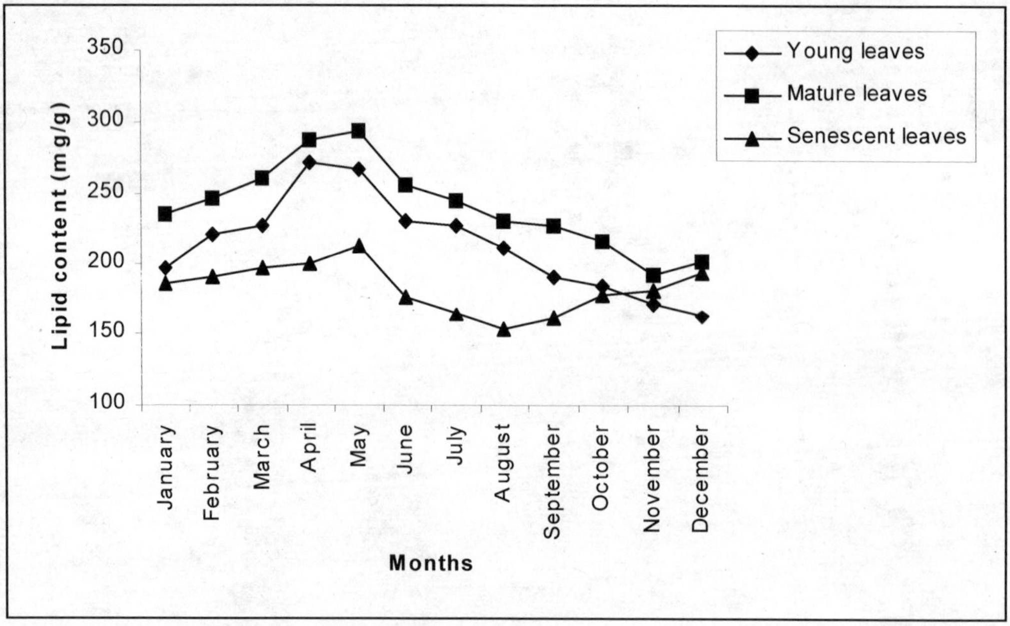

Fig. 3.4: Seasonal Fluctuation in the Level (mg/g) of Lipid Content in the Leaves of *Cestrum diurnum* in Different Physiological Stages

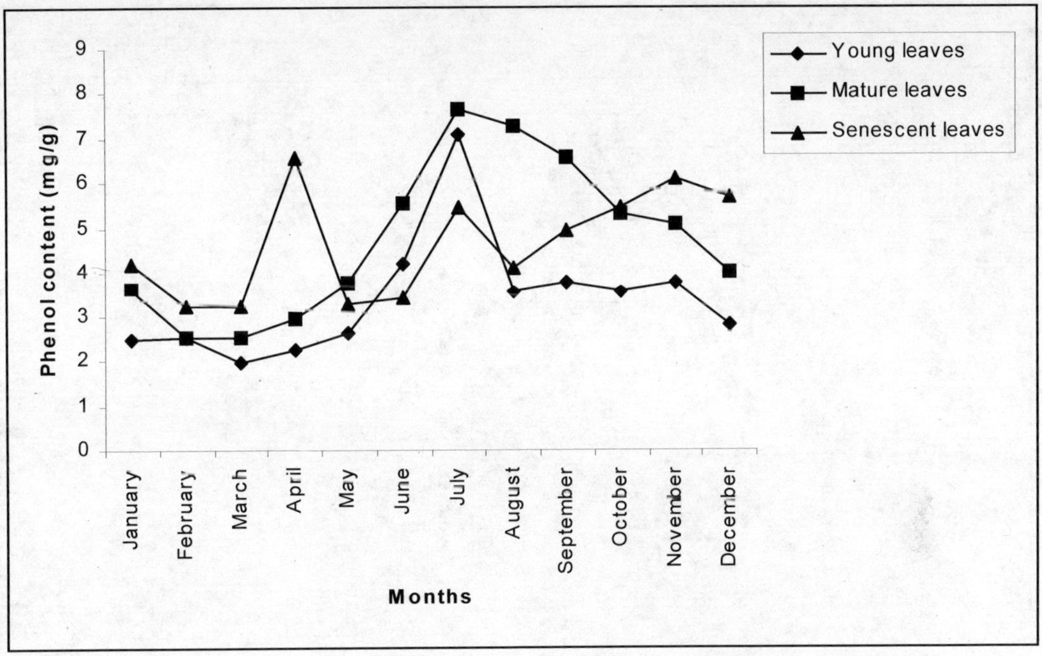

Fig. 3.5: Seasonal Fluctuation in the Level (mg/g) of Phenol Content in the Leaves of *Cestrum diurnum* in Different Physiological Stages

Ash and Moisture Contents

Seasonal changes in the moisture content revealed that the amount of moisture in mature leaves was always higher than young and senescent leaves (Fig. 3.6). The highest concentration of moisture was found in the month of July for mature and senescent leaves and in the month of August for young leaves. The lowest concentration was found in the month of January for all kind of leaves.

The pattern of seasonal changes in the ash content revealed that the higher amount was found in senescent leaves than in young and mature leaves except in the months of September–November when the highest amount of ash content was found in mature leaves (Fig. 3.7). The highest amount of ash content was found in the month of May, February and August in senescent, mature and young leaves respectively. The lowest amount was found in the month of October, May and February for senescent, mature and young leaves respectively.

Essential Oil Analysis

Gas liquid chromatographic analysis of the essential oil of mature leaves of Cestrum, diurnum (L.) revealed the presence of some saturated and unsaturated fatty acids (Fig. 3.8). Analysis of relative amount of major identified fatty acids indicated highest concentration of palmitic acid (27.62%) followed by stearic acid (4.62%) and oleic acid (3.06%).

Discussion

Current study revealed that the levels of primary and secondary biochemicals were extremely variable in the leaves of *Cestrum diurnum* with different physiological stages. Carbohydrate is the most

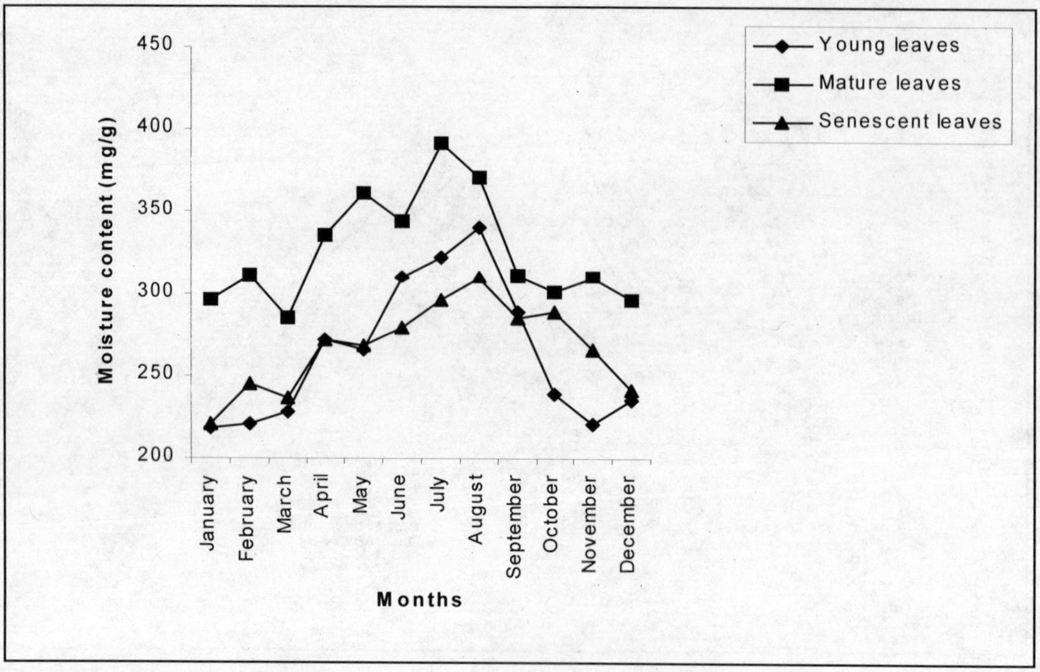

Fig. 3.6: Seasonal Fluctuation in the Level (mg/g) of Moisture Content in the Leaves of *Cestrum diurnum* in Different Physiological Stages

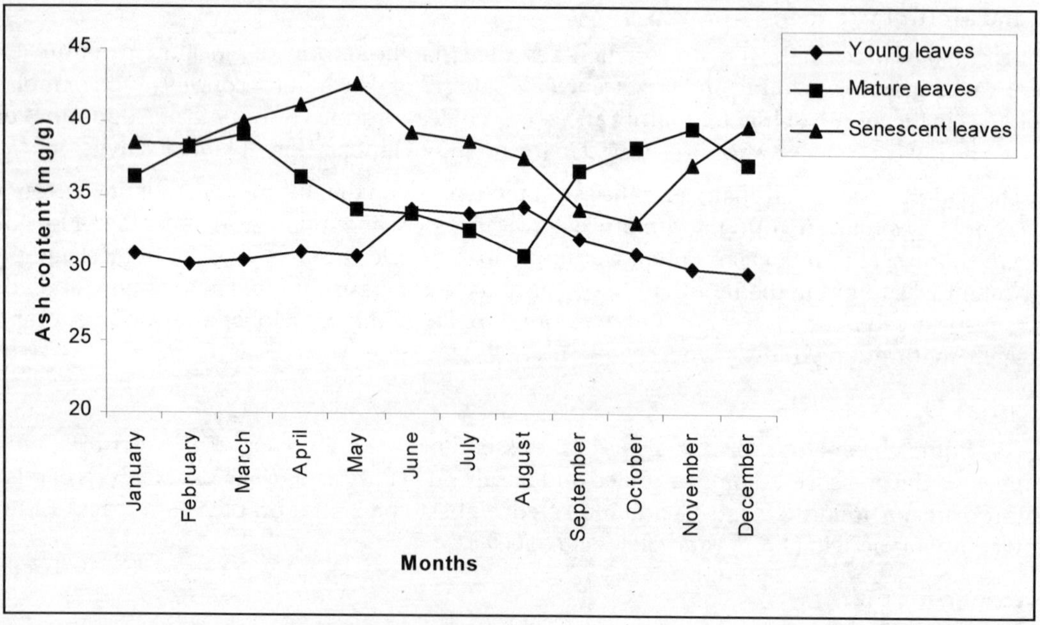

Fig. 3.7: Seasonal Fluctuation in the Level (mg/g) of Ash Content in the Leaves of *Cestrum diurnum* in Different Physiological Stages

HP 6890 Plus GLC and HP3398A GC Chemstation

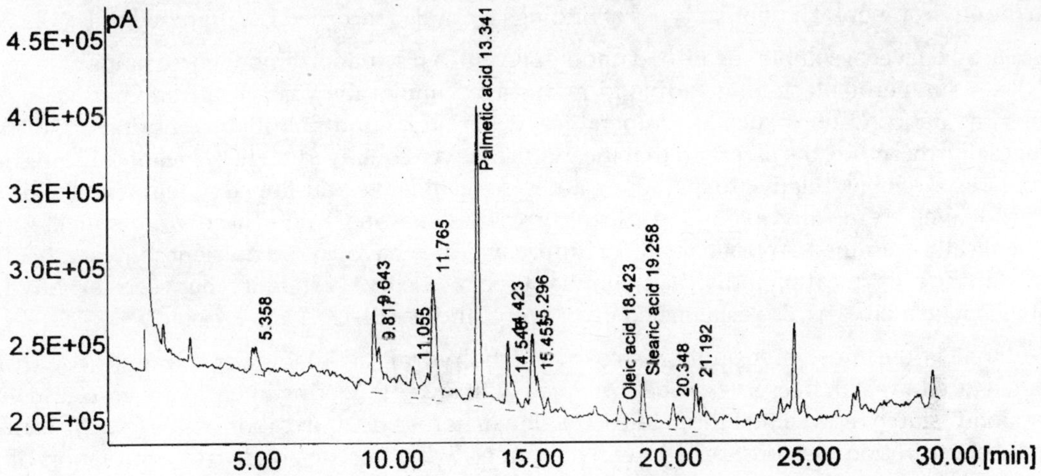

Fig. 3.8: GLC Analysis of the Essential Oil of *Cestrum diurnum*

Fig. Relative Amount of Identified and Unidentified Fatty Acid

Name of Fatty Acids	*Retention Time (Min)*	*Relative Amount*	*Area (pA*s)*
Palmetic Acid	13.341	27.62	1638291
Oleic Acid	18.423	3.06	181198
Stearic Acid	19.258	4.62	274152
Peak1	5.358	3.44	204103
Peak2	9.643	7.69	456133
Peak3	9.817	4.93	292183
Peak4	11.055	4.09	242473
Peak5	11.765	16.51	978866
Peak6	14.423	6.22	369157
Peak7	14.548	2.92	173331
Poak8	15.296	8.16	484091
Peak9	15.455	3.14	185969
Peak10	20.348	2.76	163610
Peak11	21.192	4.84	287249

important component for metabolism as it supplies the energy needed for respiration and other metabolic processes. Carbohydrate reserves are mobilized in trees to fuel maintenance respiration during winter (7), build leaves and support new growth in spring and to provide the energy expended in making adaptive responses to soil water deficits, pathogen attacks and herbivory(8-11).Starch and soluble sugars are the predominant reserve carbohydrate available for energy growth and maintenance. Current study revealed that the the proportion of photosynthecate (including soluble sugar and starch compounds) allocated leaves with different physiological stages change over the growing season and with the environment. The similar types of findings were also recorded in different plants (12-13).

Seasonal level of soluble sugars had undergone much fluctuation among three kinds of leaves of the shrub. Numerous factors including growth stage, time of the year, fertilizers, herbicides and environmental conditions such as temperature, water stress and pollution can bring about these changes. Furthermore, the assertion that the mature leaves contained higher level of carbohydrates during May-October, relative to the young and senescent leaves, sustained herein. The fact is that mature leaves were fully expanded during growing seasons when these were collected and biochemically examined. And as a result, mature leaves may have act as a nutrient source rather than a sink for the foliage-eating individuals. Similar type of positive correlation between carbohydrate content and growth was also established in different studies (14-17).

Starch is the most predominant reserve carbohydrate available for energy, growth and maintenance.Low starch reserve can be correlated with lower flowering intensity, fruit set and yield. Commonly, starch represented the greatest fraction of non-structural carbohydrate (NSC). Peak leaf starch concentration was observed in early summer, which corresponded to completion of leaf expansion and maturation. Active growth in the growing season resulted in a decline in stored starch reserve. During this time, accumulated starch is gradually converted to increased amount of sugars. Starch content decreased following shoot emergence and declined until flower bud formation. Following bud initiation levels of starch increased during flowering and this trend continued until the plant enters into senescent phase.

Protein comprises one of the most important biomacromolecule and a major determinant in controlling tree growth rate and reproduction. The protein content differed substantially amongst young, mature and senescent leaves during growing and reproductive phase of the plant (18-20). Similar to other plant systems (21-25) *C.diurnum* leaves also had higher quantity of soluble protein in mature leaves during growing season and the minimum level was found during the flowering season. This was probably due to the allocation of resource materials in the flower and fruit etc. to act as a sink rather than a source. The results exhibited here suggested that the soluble protein could be used as an indicator of the physiological state of the plant.

The richness in secondary compounds in plants is, at least, partly explicable by the simple fact that plants are rooted in the soil and cannot move; they cannot respond to the environment in ways open to animals. Theories of plant chemical defenses discriminate between quantitative digestibility reducers such as, tannins and phenolics and qualitative toxin such as, steroids and alkaloids.

Phenolics are thought to be the part of the natural defense mechanism of the plant to prevent severe defoliation. The poor livestock performance coincides with a high level of phenolics in the leaf. Present study revealed that the amount of phenolics exhibits an inverse relationship with the total protein contents in early growing season (June to August). Similar relationship has also been observed in several species of vascular plants (26-27) and macro algae (28-29). When nitrogen limits growth, there may be more fixed carbon that can be metabolized into proteins, thus plants accumulate phenolics or other nitrogen free constituents.

Thin layer chromatographic analysis revealed the presence of steroid and alkaloid compounds in the young, mature and senescent leaves (with distinct Rf respectively) may be concluded as the herbivore grazing inhibitor as well as responsible for the calcinogenic activity of the plant.

Estimation of the moisture content and ash content is important to determine the values of other compounds in the leaves. High moisture content can cause over heating of the leaves and deterioration of quality, particularly increase in free fatty acid contents; whereas high ash content is positively related to plant senescence due to an increase in the total crude fibers.

Lipids are unique class of chemical compounds that all organisms need to survive. They are primarily used for energy storage, membrane structure and as hormones. Seasonal variations in lipid content, with the maximum in spring, were observed at all sampling stations. Most of this lipid was tightly bound to the wall fabric. Temperature has a characteristic effect on many plant lipids in which it increases the level of instauration of acyl chains, which slows down both metabolism and transport. Therefore, the highest lipid content of *Cestrum diurnum*. was found at the beginning of the growing season *i.e.* in spring during the study period. Consequently, as the temperature increased, the lipid level decreased and remained almost stable until the end of the growing season. The functional units of lipids are fatty acids. The discovery in the mid-1950's of gas-liquid chromatography (GLC) has revolutionized the analysis of fatty acids and, undoubtedly, this technique is the most frequently used, for the quantification of individual fatty acid in any acylated lipids. The present study indicates a high percentage of saturated fatty acid in comparison to unsaturated fatty acids. This is probably due to the distribution pattern of *Cestrum diurnum*, available in the tropical rather than in temperate zone.

REFERENCES

1. Little, E.L., Jr., R.O. Woodbury, and F.H. Wadsworth. (1974) Trees of Puerto Rico and the Virgin Islands. Vol. 2. Agric. Handb. 449. U.S. Department of Agriculture.
2. Project report-Calcinogenic activity of *Cestrum diurnum*,Animal science division Colonel University,2001.
3. Dubois, M., Gilles, K.A., Hamilton, J.K., Pebers,P.A. and Smith,F.(1956) Colorimetric method for determination of sugar and related substances. Analytical Biochemistry. (28) 350-356.
4. Lowry, O.H., Rosebrough, N. J., Farr,A.L. and Randal,R. J. (1951) Protein measurement with folin phenol reagent. Journal of Biological Chemistry (193). 265-275.
5. Folch,J,Lees, M and Solane-stanley,G.H.(1957). A simple method for the isolation and purification of total lipids from animal tissues.J.Biol.Chem.(226),497-507.
6. Bray,H.G. and Thrope,W.V.(1954). Analysis of phenolic compounds of interest in metabolism.Meth.Biochem.Anal (1) 27-52.
7. Ögren, E. (2000) Maintenance respiration correlates with sugar but not nitrogen concentration in dormant plants. Physiol. Plant (108)295–299.
8. Wargo, P.M. (1981) Defoliation and tree growth. *In* The Gypsy Moth:Research Toward Integrated Pest Management. Eds. C.C. Doane and M.L. McManus. Tech. Bull. (1584) USDA, Washington,pp 225–248.
9. Oren, R.. Schulze, E.D. Werk, K.S. Meyer J. Schneider, B.U. and. Heilmeier, H (1988) Performance of two *Picea abies* (L.) Karst.stands at different stages of decline. I. Carbon relations and standgrowth. Oecologia (75)25–37.
10. Dunn, J.P., D.A. Potter and T.W. Kimmerer.(1990) Carbohydrate re-serves, radial growth, and mechanisms of resistance of oak trees to phloem-boring insects. Oecologia (83)458–468.
11. Canham, C.D., R.K. Kobe., E.F. Latty and R.L. Chazdon.(1999) Inter-specific and intraspecific variation in tree seedling survival: effectsof allocation to roots versus carbohydrate reserves. Oecologia.(121)1–11.

12. Kramer, P.J. and. Kozlowski, T.T. (1979). Physiology of woody plants.Academic Press, London, 811 p.

13. Terziev, N., Boutelje J.and Larsson K. (1997) Seasonal fluctuations of low-molecular-weight sugars, starch and nitrogen in sapwood of *Pinus sylvestris* L. Scand. J. For. Res. (12)216–224.

14. Fischer, C. and Höll. W. (1992) Food reserves of scots pine (*Pinussy lvestris* L.). II. Seasonal changes and radial distribution of carbo-hydrate and fat reserves in pine wood. Trees (6)147–155.

15. Höll, W. (1997) Storage and mobilization of carbohydrates and lipids. *In* Trees-Contributions to Modern Tree Physiology. Eds. H. Ren-nenberg, W. Eschrich and H. Ziegler.Backhuys Publishers,Leiden, The Netherlands, pp 197–211.

16. Priestley, C.A. (1960) Seasonal changes in the carbohydrate resources of some six-year-old apple trees. East Malling Res. Sta. Annu. Rep.(1959) pp 70–77.

17. Saranpää, P. and Höll, W.(1989). Soluble carbohydrates of *Pinus sylvestris* L. sapwood and heartwood. Trees (3)138–143.

18. Rao, C.H.K.; Chauhan,V.D. and Indusekhar, V.K. Elemental (1983) Composition of certain green seaweeds of Porbandar reef. Indian Journal Marine Sciences. (12) 197-199.

19. Fadeel, A.A., Metwalli,A., and Al-houty, F.A. (1987) The effect of seasonal variation in temperature and light intensity on the nitrogenous contents in four species of the chlorophyceae of Kuwait. Dirasat(VIX) 107-118.

20. Ochiai, Y., Katsuragi T. and Hashimoto, K. 1987 Proteins in three seaweeds "aosa" Ulva lactuca, "arame" Eisenia bicyclis, and "makusa"Gelidium amansii. Nippon Suisan Gakkaishi (53) 1051-1055.

21. Munda,I. M., Gubensek F. (1986) The amino acid composition of some benthic marine algae from the Northern Adriatic. Bot. Mar. (29) 367–372.

22. Munda I. M., Gubensek F. (1976) The amino acid composition of some common marine algae from Iceland, Bot. Mar (19) 85–92.

23. Owens, N. J. P., Stewart,W. D. (1983) Enteromorpha and the cycling of nitrogen in a small estuary, Estuar. Coast. Shelf Sci 17 (3), 287–296.

24. Tkachenko F. P., Koval V. T. (1990) Biochemical composition of abundant benthic seaweeds of the Black Sea, Hydrobiol. J., 26 (6), 39–43.

25. Wheeler P. A., Bj¨orns¨ater B. R. (1992) Seasonal fluctuations in tissue nitrogen, phosphorus and N:P for five macroalgal species common in the Pacific northwest coast, J. Phycol. (28) 1–6.

26. Mattson, WJ. (1980)Herbivory in relation to plant nitrogen content.Ann.Rev. Ecol. Syst. (11) 119-161.

27. Buchsbaum, R.F.T., Short and D.P.Cheney.(1990)Phenolic-nitrogen interactions in eelgrass,Zostera marina L.:possible implications for disease resistance.**Aquat.Bot**.(37)291-297.

28. Neish.Shacklock, A.P. Foc, C and.Simpson F (1977) The cultivation of Chondrus crispus.Factors affecting growth under greenhouse conditions.**Can.J.Bot** (55) 2263-2271.

29. MaclerB. (1986) Regulation of carbon flow by nitrogen and light in red algae, *Gelidium coulteri*.Plant Physiol.(82)136-141.

Advances in Biochemistry and Biotechnology Volume 1 *Pages 95-106*
Edited by: **Dr. Chiranjib Chakraborty**
Published by: **DAYA PUBLISHING HOUSE**

4

Protein Folding Pathways: Its Kinetics and Related Functions

*Surajit Sinha, Shyam Sundar Nandi, Akash Verma, Ragini Gothalwal & T.K.Maiti**
**Dept. of Biotechnology, I.I.T.-Kharagpur (W.B)*
E-Mail: tkmaiti@hijli.iitkgp.ernet.in

ABSTRACT

Among the nanobiomolecules, proteins are the most mysterious because of their hugely diversified functions ranging from cellular growth, cell-cell interactions to cellular signaling which are governed by their biologically active 3D structure. Moreover, proteins function in a fluctuating cellular environment, stabilized by disulphide bonds, which influence the thermodynamics of protein folding. Proteins that misfold and exhibit decreased solubility under physiological conditions results in protein disorders such as Parkinson's and Alzheimer's disease. These set of disorders are characterized by either local or systemic extracellular tissue deposition of specific proteins under physiologically attainable conditions. This deposition can lead to a compromise in organ function, which in most cases, is fatal without treatment. Innovative use of

* Correspondent Author

computational methods along with experimental data are thus needed to understand the rate and kinetics of protein folding and protein folding pathways which can provide insights in to the basis of such protein disorders and help develop novel therapeutic approaches.

Keywords: *Nanobiomolecules; kinetics; protein refolding; breaker peptides.*

Introduction

Protein function is a measure of protein stability and protein structure. Proteins are the center of the action in biological processes and their function can only be known based on the structure of their constituent polypeptide chains. Therefore, understanding the protein folding mechanism (1-3) is critical in molecular biology not only because it is one of the fundamental problems remaining in protein science, but also because several fatal diseases are directly related to protein folding/misfolding, such as the Alzheimer's disease, Mad cow disease, Prion disease and cystic fibrosis disease. Studies on protein folding poses two fundamental problems. One is the Levinthal's Paradox which refers to how proteins find their correctly folded state on a non-astronomical time scale given the myriads of possible conformations (4,5). Second, how this efficient process of protein folding is carried out in vivo taking in to account the adverse conditions in the crowded environment of the cell. The failure of molecular chaperones (6-8) in preventing aggregation in these fatal cases is still a question to be answered. Despite the enormous efforts made by various groups, the problem of protein folding still remains largely unsolved. Experiments that probe proteins at different stages of the folding process have helped to elucidate kinetic mechanisms and the thermodynamics stabilities of folding (9-11). For example- the observation that early stages of protein folding involves single-exponential kinetic phases, can be interpreted as evidence that these phases are rate limited by significant energy or entropy barriers. Computer stimulation (12,13) performed at various levels of complexity ranging from simple lattice models to all atom models with explicit solvent are used to supplement experiment and fill in some of the gaps in our knowledge about folding pathways. Because explicit solvent stimulations typically require enormous CPU time, many recent studies have been carried out with implicit solvent models. However, it is still largely an open question as to how well these implicit solvent models can predict the thermodynamics as well as the kinetics of protein folding. We are not very close to realizing this goal, and so the protein-folding problem remains one of the most basic unsolved problems in computational biology. Here, we review how disulphide bonds and contact order have helped us in elucidating the protein folding pathways, its function and kinetics with special reference to protein conformation disorders (PCD). Lastly, we look at protein refolding therapy, which if possible can have enormous applications ranging from therapeutic benefits to artificial modeling of the cell.

Disulphide Bond as Probes for Conformational Transitions

The common belief is that disulphide bonds have been added during evolution to increase the stability of proteins that function in a fluctuating cellular environment (14,15). Disulphides bonds help stabilize the folded states by increasing favourable enthalpy interactions in the folded states and by lowering the entropy of the unfolded states (16,17). But recent evidence indicates that disulphide bonds are not inert structural motifs but play important role in protein conformation and function. Not only because the interaction among thiol groups of cysteine residues are of reductive/oxidative nature and therefore requires the presence of an appropriate electron acceptor or donor, which can be controlled experimentally as can the intrinsic stability of the disulphide bond interaction but because of the fact that the rates of forming or breaking of protein disulphide bonds reflects the protein conformational transitions that occur. Therefore, the equilibrium between folded and unfolded conformations can be altered simply by changing only the thiol/disulphide intrinsic stability (18-20). As a result, all disulphide bonds present in a protein at any time may be trapped in a stable form and be chemically identified. A number of different ways of forming and breaking protein disulfides are known, but the most suitable for studies of protein folding involves the use of thiol-disulphide exchange between the protein and a small molecule reagent in either thiol or disulphide form. The most commonly used reagent is glutathione (GSSG).

The disulfides present at any instant in a protein can be trapped in a stable form by simply blocking, rapidly and irreversibly, all thiol groups in the solution with the help of a quenching reagent (Fig. 4.1). Therefore, in the absence of other nucleophiles, disulfide bonds may no longer be made, broken, or interchanged. The quenching reaction should be very rapid so that it faithfully traps the species present at the time of addition of the quenching reagent. Trapping reactions with reagents like iodoacetamide or iodoacetic acid has the advantage of being irreversible, but the reaction with a free thiol group has a half-time of 0.3 seconds with 0.2M iodoacetate at pH 8.7 and 25°C. Therefore, disulphide rearrangements could occur during the trapping reaction. The other problem being the fact that buried thiol groups do not react at a sufficient rate during the trapping procedure. Nevertheless,

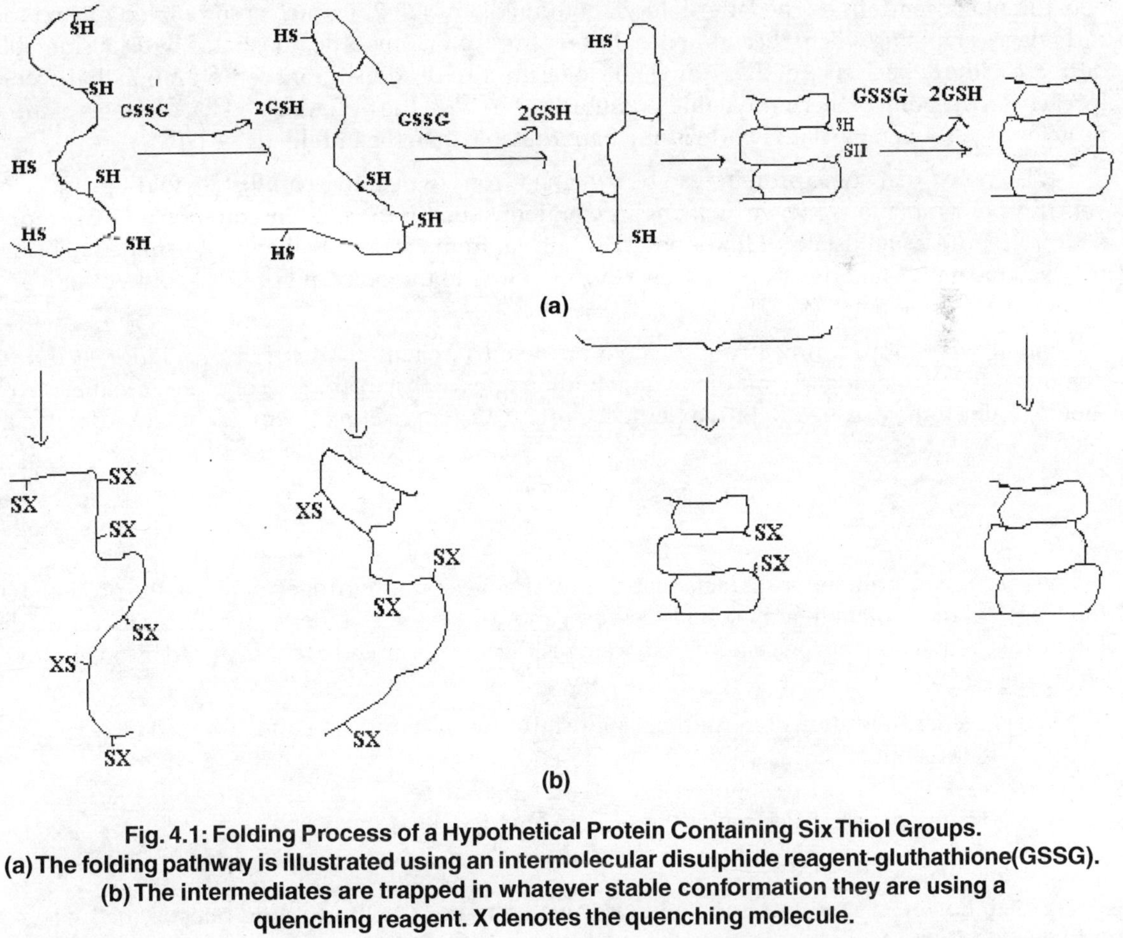

Fig. 4.1: Folding Process of a Hypothetical Protein Containing Six Thiol Groups.
(a) The folding pathway is illustrated using an intermolecular disulphide reagent-gluthathione(GSSG).
(b) The intermediates are trapped in whatever stable conformation they are using a quenching reagent. X denotes the quenching molecule.

once all free thiol groups are blocked irreversibly, the trapped species are stable indefinitely. They can then be separated and characterized in detail. The intermediate species trapped by either acid, iodoacetate or iodoacetamide can be analyzed by nondenaturing polyacrylamide gel electrophoresis, ion-exchange chromatography and by reversed-phase HPLC. Since, the quenching reagents affect the equilibrium between folded and unfolded conformations and the rates of disulfide bond formation, the kinetics of unfolding and refolding can be determined unambiguously under all unfolding and refolding conditions. Some of the proteins whose folding pathways have been elucidated using disulfide bonds includes Bovine Pancreatic Trypsin Inhibitor (BPTI), dendrotoxins I and K from the venom of black mamba snakes, Ribonuclease T_1 and Ribonuclease A.

Kinetics and Contact Order

The ultimate goal of kinetic folding studies is the molecular elucidation of the successive structure formation that takes place during folding. But there is a significant difference in the folding behavior of small proteins with simple two-state kinetics, and of larger proteins having a three-state folding kinetics (21,22). The former have no visible intermediates in the course of folding, which therefore occurs as an " all-or-none " process under all experimental conditions. This peculiarity is demonstrated by linear dependence of logarithms of the folding (and unfolding) rates of these proteins on the denaturant concentration. The latter fold via intermediates (23-25) which accumulated during the early stages of folding when it occurs in denaturant- free water. This is demonstrated by a considerable decrease in the dependence of the folding rate logarithm on the denaturant concentration that occurs far off from the point of thermodynamic equilibrium between the unfolded and native states of these proteins, that's when the native form is much more stable than the unfolded one (26,27).

It is known that some proteins can be switched from two-state to multistate folding, and vice versa, by point mutations or even by changing conditions such as the salt concentration or temperature (28,29). In addition multistate folding is observed only far from the point of thermodynamic equilibrium between the native and denatured states, whereas, close to the point, all proteins fold without any observable intermediates (30-32).

Small two-state folding proteins have attracted particular attention of scientists. It was demonstrated that the logarithms of in-water folding rates of these proteins correlate with their gross topological parameter called **Relative Contact Order** (CO: Plaxco *et al.*, 1998) (33-35). CO is defined as

$$CO = \frac{1}{L \cdot N} \sum^{N} \Delta L_{ij}$$

where N is the number of contacts (within 6 A^0) between nonhydrogen atoms in the protein, L is the length of the protein in amino acid residues, and ΔL_{ij} is the number of residues separating the interacting pair of nonhydrogen atoms (adjacent residues are assumed to be separated by one residue, etc.).

CO is a renormalization of the perhaps more intuitive measure, **Absolute Contact Order**,

$$Abs\_CO = \frac{1}{N} \sum^{N} \Delta L_{ij} = CO \times L$$

Which however, was found to be less correlated than CO with folding rates of the two-state folders (36). The CO was invented to compare differences in topology (rather than in size) between proteins of different length (37). This parameter is small for proteins stabilized mainly by local interactions and is large when residues in a protein interact frequently with partners far away in the protein sequence.

However, CO parameter cannot predict the folding rates of proteins with three state kinetics and for short peptides. Instead, the three state folding proteins demonstrate a strong anticorelation between their size and the folding rate logarithm. It seems likely that CO takes in to account topology only and plays no explicit attention to the protein size. Therefore, a much more specific approach is needed for understanding the kinetics of three state folders.

Energy Profiling

The kinetic reaction involving intermediates in a protein folding process is generally represented as

$$U \xrightarrow{K_{UI}} I \xrightarrow{K_{IN}} N$$

Where the formation of an intermediate (I) causes a lag in the appearance of the product (N-native). The magnitude of the lag period depends upon the ratio of the two rate constants. Spectroscopic techniques are usually not well suited to detect such lag phases since intermediates are often intermediate in their spectral properties between U and N (38). The lag that would be observed in the I→N reaction is compensated by the contribution of the U→I reaction and thus is not visible in the kinetics. The formation of native protein molecules via intermediates during folding or refolding can be measured selectively by a technique that is related to the energetic profile of folding (Fig. 4.2). The completely folded species, N, is separated from all other unfolded or partially folded species by a high activation barrier. Therefore, only N molecules unfold slowly, whereas all intermediates that have not yet crossed this final barrier unfold rapidly. Therefore the formation of native protein can be measured by unfolding assays. Folding is initiated inthese experiments by dilution to the desired refolding conditions at time t − 0 in a test tube. Then, after various time intervals of refolding, samples are withdrawn and are transferred to unfolding conditions. The amplitude of the resulting unfolding reaction (by absorbance or by fluorescence) is a measure of the amount of completely refolded molecules that were present at the time when folding was interrupted. The maximum value of the unfolding amplitude is obtained after folding in the first step of the assay has gone to completion and should correspond to the amplitude that is observed when native protein is unfolded under the same conditions. The folding kinetics of great many proteins have been determined in this way including Rnase and α-lactoglobulin. However, many of the details of protein folding pathways including the structure of the intermediates remain unknown and as long as these are not solved treatment of diseases arising out of protein folding/misfolding disorder will be far from reality.

Protein Aggregation and Transition

The biological function of a protein depends on its three dimensional structure, determined by its amino acid sequence during the process of protein folding. Diverse diseases have been shown to arise from protein misfolding and a number of misfolded proteins and peptides aggregate in to insoluble fibrils (Fig. 4.3) (38,39). The aggregates are toxic and can poison the cells around it. The aggregation of these diseases are related to fatal diseases (40,41). That protein conformation plays an essential role in the development of pathological consequences for the organism became clear in the course of studying the molecular basis of diseases like Prion disease, Alzheimer's disease, Cystic Fibrosis, Mad Cow

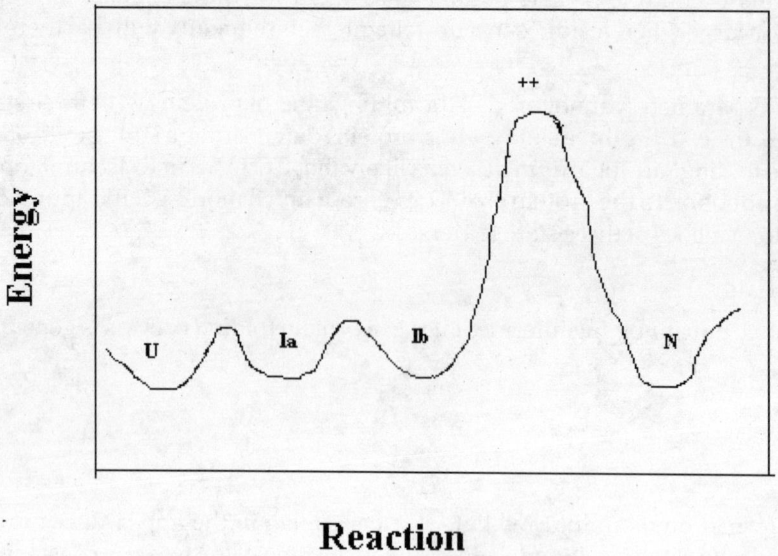

Fig. 4.2: Energy Profile for a Protein Folding Reaction.
++ Denotes activated state close to the native structure N. U is unfolded protein; Ia and Ib are partially folded intermediates.

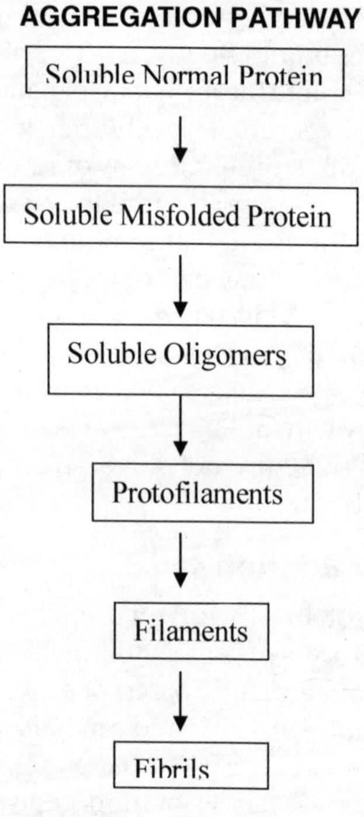

Fig. 4.3: Pathway of Fibril Formation

disease, Parkinson's disease, transmissible spongiform encephalopathies (TSE), Amyotrophic lateral sclerosis (ALS) and others. All these diseases are grouped under the name Protein Conformational Disorders (PCD) (Table 4.1). Diorders of such nature are classified by the accumulation of protein aggregates at specific sites for that particular protein; neurodegenerative diseases, such as Parkinson's or Huntington's disease, form intracellular aggregates, and the systemic amyloidoses form extracellular aggregates. The amyloidoses were the first recognized examples of protein deposition diseases. These deposits bind the dyes thioflavin-S,-T and congo red, the latter with positive birefringence under polarized light (42,43). In Alzheimer's disease, the prion disorders and the islet amyloid polypeptide deposits in pancreatic islets of the elderly, deposition is local, the modes of pathogenesis are less clear, and there is evidence for intracellular aggregation as well as extracellular deposition (44). Electron microscopy of the deposits reveal a discrete fibrillar structure. Recently, proteins not implicated in amyloid diseases were found to form fibril structures in vitro under denaturing conditions (45,46), suggesting that fibril formation is a common feature of destabilized proteins(28). Regardless of sequence and structures of proteins, the fibrils have similar core structures, mainly composed of β-sheets(47). For example, the native Aβ peptide in Alzheimer's disease(48,49) and prion proteins (PrP^c) in prion diseases(50,51) are α–helix rich. The aggregation of α-helix rich proteins or peptides involves a conformational transition from α-helices to β-sheets. A similar transition has also been observed in vitro in some α-helical peptides (52,53) that aggregate in to amyloid fibrils by means of changing the environment, such as varying the organic solvent condition (54), altering the pH, and controlling the redox state. Moreover, similar α-β transitions also occur through the correct folding pathway in proteins with a nonhierarchical folding mechanisms. For example, β-lactoglobulin, a predominant β-sheet protein (55) is observed to form non-native α-helical intermediates on folding. Thus understanding the α-β transition is important for both protein folding and protein aggregation. The secondary structures of proteins, mainly α-helices and β-sheets, are determined by the amino acid sequence and stabilized by hydrogen bonds. However, under denaturing conditions proteins with various organizations of the secondary structure elements can aggregate in to similar β-rich amyloid fibrils. This suggests that sequence non-specific hydrogen bond interaction among the backbone of proteins is an important factor for aggregation (56).

Protein Refolding Therapy

Since the 1950, scientists knew that the information necessary for a protein to fold is encoded in its aminoacid sequence. But do proteins fold by themselves or do they need some machines which assists them in folding?. It became clear only during the late sixties, that cellular protein folding critically depends on the assistance of so called chaperones (57-60). This specialized set of proteins is able to prevent the aggregation of nascent or misfolded substrates by providing microenvironments to protect them from nonproductive interactions (61). These proteins have hydrophobic surfaces that recognizes and bind to exposed hydrophobic surfaces of improperly folded proteins and prevents non-native interactions and thereby aggregation. It is this very concept that has been utilized in targeting refolding therapy. Considering the fact that protein misfolding and aggregation are central in the pathogenesis of PCD, any therapy directed to the cause of the disease, therefore, should aim to inhibit or reverse the conformational changes that result in the formation of the pathological protein conformer. Since, protein misfolding is triggered by conformational changes stabilized by oligomerization, the strategy should be to prevent the stabilization of the misfolding or to destabilize the monomeric intermediates and the early β-sheet oligomers (62,63).

It has been postulated that short synthetic peptides containing self-recognition motif of the protein and engineered to destabilize the abnormal conformation might be useful to correct protein folding.

These peptides are called as synthetic mini-chaperones. These are designed to be similar to the sequence of the protein region responsible for oligomerization and contain residues that specifically favor or disfavor a particular structural motif. Infact such a mini-chaperone known as β-sheet breaker peptide have been designed and has been found to block the conformational changes and aggregation in both Alzheimer's disease (Aβ-peptide) and in Prion disease (PrP^c –prion protein) both *in vitro* and *in vivo*. These results suggests that these synthetic β-sheet breaker peptides or synthetic minichaperones could be used in destabilizing the β-sheet rich abnormal conformation and inducing its conversion in to the normal form (62). These β-sheets breaker peptides are short, synthetic peptides, homologous to the sequence of the protein region thought to play a central role in the conformational change, and containing amino acid residues that disfavor the adoption of a β-sheet conformation. Studies *in vitro*, in cell cultures and *in vivo*, have shown that β-sheet breaker peptides, homologous to the PrP fragment which contain proline residues to destabilize β-sheet structures, were able to prevent and reverse the PrP^C ? PrP^{Sc} (64-66). Also under investigation is whether molecular chaperones–the cell's own chaperones –can be used to fix protein misfolding problems. It had been suggested that drugs that could boost chaperone activity could do the trick. Recent findings that geldanamycin treatment protected flies against dopaminergic neuron loss had laid a general platform technology to design drugs for treatment of PCD.

Table 4.1: List of Some Protein Disorders Associated with Protein Misfolding

Protein Involved	*Associated Diseases*
β-Amyloid	Alzheimer's Disease
α- Synuclein	Parkinson's Disease
Huntingtin	Huntington's Disease
Prion Protein(PrP^c)	Prion Disease
CFTR Protein	Cystic Fibrosis
Hemoglobin	Sickle cell anemia
SOD	Amyotrophic lateral sclerosis
Amyloid- A	Reactive amyloidosis

Conclusion

Solving the folding problem has enormous applications: exact drugs can be designed theoretically on a computer without a great deal of experimentation. Genetic engineering experiments to improve the function of particular proteins will be possible. Moreover, stimulating protein folding can allow us to go forward with the modeling of the cell. Catching hold of intermediates in protein folding pathway can help in designing synthetic mini-chaperones that can prevent proteins proceeding through such intermediates and finally aggregating. Therefore, understanding the protein folding pathways can help us develop principles for generating synthetic mini-chaperones that could be used as novel therapeutic approach for disorders where a function of protein needs to be modified. Also, since the phenotypic characteristics of living organisms depend upon the integrative function of thousands of proteins, engineering mini-chaperones that could alter the structure of specific protein could enable us manipulate phenotypic features of living organisms and could reveal a new therapeutic strategy for a variety of diseases.

REFERENCES

1. Dobson, D.M., Sali,A., and Karplus, M. (1998). Protein Folding: A Perspective from theory and experiment. **Angew. Chem. Int. Ed.** (37),868-893.

2. Ali, A. (1998). 100,000 protein structures for the biologist. ***Nature Struct. Biol.*** (5) 1029-1032.

3. Karplus, M., and Ali, A. (1995). Theoritical studies of protein folding and unfolding. **Curr. Opin. Struct. Biol**. (5) 58-73.

4. Gutsche, I., Essen, O. L., and Baumeister, W. (1999). Group II Chaperonins: New Tric(k)s and Turns of a Protein Folding Machine. **J. Mol. Biol**. (293) 295-312.

5. Callender, R.H., Dyer, R.B., Gilmanshin, R., *et al.* (1998). Fast events in protein folding: the time evolution of primary processes. **Annu. Rev. Phys. Chem**. (49) 173-202.

6. Frydman, J., Nimmesgern, E., Erdjument-Bromage., *et al.* (1996). Principles of chaperone-assisted protein folding: differences between *in vitro* and *in vivo* mechanisms. **Science** (272) 1497-1502.

7. Frydman, J., Nimmesgern, E., Ohtsuka, K., *et al.* (1994). Folding of nascent polypeptide chains in a high molecular mass assembly with molecular chaperones. ***Nature*** (370) 111-117.

8. Netzer, WJ. and Hartl, FU. (1998). Protein Folding in the cytosol: chaperonin-dependent and independent mechanisms. **Trends Biochem. Sci**. (23) 68-73.

9. Ferrara, P., Apostolakis, J. and Caflisch, A. (2000). Thermodynamics and kinetics of folding of two model peptides investigated by molecular dyanamics stimulations. **J Phys Chem** (104) 5000-5010.

10. Daggett, V. and Levitt, M. (1992). Molecular dyanamics stimulations of helix denaturation. ***J. Mol. Biol***. (228) 1121-1138.

11. Borreguero, J.M., Dokholyan, N.V., Buldyrev, S.V., *et al.* (2002). Thermodynamics and folding kinetic analysis of the SH3 domain from discrete molecular dynamics. **J. Mol. Biol**. (318) 863-876.

12. Takada, S., Luthey-Schulten, Z. and Wolynes, P.G. (1999). Folding dynamics with nonadditive forces: a stimulation study of a designed helical protein and a random heteropolymer. **J. Chem. Phys**. (110) 11616-11629.

13. Dokholyan, N.V., Buldyrev, S.V., Stanley, H.E., *et al.* (2000). Identifying the protein folding nucleus using molecular dynamics. **J. Mol. Biol**. (296) 1183-1188.

14. Buxbaum, N.J. (2003). Diseases of protein conformation: what do in vitro experiments tell us about in vivo diseases?.**Trends in Biochemical Sciences** (28) No. 1, 585-592.

15. Hogg, J.P. (2003). Disulphide bonds as switches for protein function. **Trends in Biochemical Sciences** (28) No. 4, 210-214.

16. Chuang, C.C., Chen, Y.C., Yang, M.J., *et al.* (2003). Relationship between protein structures and disulphide bonding patterns. **Proteins** (52) 1-5.

17. Abkevich, V.I. and Shaknovich, E.I. (2000). What can disulphide bond tell us about protein energetics, function and folding: stimulations and bioinformatics analysis. **J. Mol. Biol.** (*300*)975-985.

18. Clarke, J., Hounslow, A.H., Bond, C.J., *et al.* (2000). The effects of disulphide bonds on the denatured state of barnase. **Protein Sci** (9)2394-2404.

19. Cavagnero, S., Dyson, H.J. and Wright, P.E. (1999). Effect of helix destabilizing mutations on the kinetic equilibrium folding of apomyoglobin. **J. Mol. Biol**. (285) 269-282.

20. Laurents, D.V., Corrales, S., Elias-Arnanz, M., *et al.* (2000). Folding kinetics of phage 434 Cro protein. **Biochemistry** (39), 13963-13973.

21. Oxana, V.G., Garbuzynskiy, O.S. and Ivankov, N.D. (2003). Chain length is the main determinant of the folding rate for proteins with three-state folding kinetics. **Proteins** (51) 162-166.

22. Fersht, A. R. (2000). Transition-state structure as a unifyingbasis in protein-folding mechanisms: contact order, chain topology, stability, and the extended nucleus mechanism. **Proc. Natl. Acad. Sci**. *USA* (97) 1525-1529.

23. Eaton, W.A., Munoz, V., Hagen, S.J., *et al.* (2000). Fast kinetics and mechanisms in protein folding. **Annu. Rev. Biophys. Biomol. Struct**. (29) 327-359.

24. Plaxco, K.W., Simmons, K.T., Baker, D., *et al.* (1998). Contact order, transition state placement and the refolding rates of single domain proteins. **J. Mol. Biol.** (227) 985-994.

25. Thirumalai, D. (1995). From minimal models to real proteins: Time scales for protein folding kinetics. **J. Phys.** (5) 1457-1467.

26. Galzitskaya, O.V., Ivankov, D.N. and Finkelstein, A.V. (2001). Folding nuclei proteins. **FEBS Lett,** (489) 113-118.

27. Finkelstein, A.V. and Badretdinov, A.Y. (1997). Rate of protein folding near the point of thermodynamic equilibrium between the coil and the most stable chain fold. **Des.** (2) 115-121.

28. Ivankov, N.D., Garbuzynskiy, O.S., Alm, E., *et al.* (2003). Contact order revisited:Influence of protein size on the folding rate. **Protein Science** (12) 2057-2062.

29. Burns, L.L., Dalessio, P.M., and Ropson I.J. (1998). Folding mechanisms of three similar b-sheet proteins. **Proteins** (33) 107-118.

30. Cavegnero, S., Dyson H.J., and Wright, P.E. (1999). Effect of helix destabilizing mutations on the kinetic and equilibrium folding of apomyoglobin. **J. Mol. Biol**. (*285*) 269-282.

31. Jackson, S.E. and Fersht, A.R. (1991). Evidence for a two-state transition. **Biochemistry** (30) 10428-10435.

32. Baker, D. (2000). A surprising simplicity to protein folding. **Nature** (*405*) 39-42.

33. Stickle, D.F., Presta, L.G., Dill, K.A., and *et al.* (1992). Hydrogen Bonding in globular proteins. **J. Mol. Biol**. (226) 1143-1159.

34. Chan, H.S. (1998). Matching speed and locality. **Nature** (392) 761-763.

35. Plaxco, K.W., Simons, K.T., Ruczinski, I., *et al.* (2000). Topology, stability, sequence, and length: defining the determinants of two-state protein folding kinetics. **Biochemistry** (39) 11177-11183.

36. Kaya, H. and Chan, H.S. (2003). Simple two-state protein folding kinetics requires near –Levinthal thermodynamic cooperativity. **Proteins** (52) 510-523.

37. Huseyin. K. and Hue, S.C. (2003). Contact order dependent protein folding rates: Kinetic consequences of a cooperativity interplay between favourable nonlocal interactions and local conformational preferences. **Proteins** (52) 524-533

38. Ding, F., Borreguero, M.J., Buldyrey, V.S., *et al.* (2003). Mechanism for the a-Helix to b-Hairpin

Transition. **Proteins** (*53*) 220-228.

39. Kelly, J.W.(1998). The alternative conformations of amyloidogenic proteins and their multi-step assembly pathways. **Curr. Opin. Struct. Biol**. (*8*)101-106.

40. Tan, S.Y. and Pepys, M.B. (1994). Amyloidosis. **Histopathology** (*25*) 403-414.

41. Westermark, P. *et al.* (2002). Amyloid fibril protein nomenclature. ***Amyloid*** (*9*) 197-200.

42. Buxbaum, J.N. and Tagoe, C.E. (2002). The genetics of amyloidoses. *Annu. Rev.* **Med.** (51) 543-569.

43. Buxbaum, J.N. (2003). Diseases of protein conformation: what do in vitro experiments tell us about in vivo diseases?. **Trends in Biochemical Sciences** (*28*) No.11, 585-592.

44. Guijarro, J.I., Sunde, M., Jones, J.A., *et al.* (1998). Amyloid fibril formation by an SH3 domain. **PNAS** (*95*) 4224-4228.

45. Chiti, F., Webster, P., Taddei, N., *et al.* (1999). Designing conditions for in vitro formation of amyloid protofilaments and fibrils. **PNAS** (*96*) 3590-3594.

46. Dobson, C.M. (1999). Protein misfolding, evolution and disease. **Trends in Biochemical Sciences** (*24*) 329-332.

47. Sunde, M., Serpell, L.C., Bartlam, M., *et al.* (1997). The common core structure of amyloid fibrils by synchrotron X-ray diffraction. **J. Mol. Bio.** (273) 729-739.

48. Selkoe, D.J. (1996). Amyloid beta-protein and the genetics of Alzheimer's disease. **J. Biol. Chem.** (271) 18295-18298.

49. Lansbury, P.T. (1996). A reductionist view of Alzheimer's disease. **Acc.Chem. Res.** (*29*) 317-321.

50. Prusiner, S.B. (1997). Prion diseases and the BSE crisis. **Science** (*278*) 245-251.

51. Harrison, P.M., Bammorough, P., Daggett, V. *et al.* (1998). The Prion Folding Problem. **Curr. Opin. Struct. Biol**. (7) 53-59.

52. Takahashi, Y., Ueno, A., and Mihara. H. (1998). Design of a peptide undergoing a- b structural transition and amyloid fibrillogenesis by the introduction of a hydrophobic defect. **Chem. Eur. J.** (4) 2476-2484.

53. Fezoui, Y., Hartley, D.M., Walsh, D.M., *et al.* (2000). A de novo designed helix turn-helix peptide forms nontoxic amyloid fibrils. **Nat. Struc. Biol**. (7) 1095-1099.

54. Ono, S., Kameda, N., Yoshimura, T., *et al.* (1995). Supersecondary structure with amphiphilic β strands probed by phenylanaline. ***Chem. Lett.*** (10) 965-966.

55. Hamada, D., Segawa, S., and Goto, Y. (1996). Non-native alpha-helical intermediate in the refolding of beta-lactoglobulin, a predominantly beta-sheet protein. **Nat. Struct. Biol.** (3) 868-873.

56. Ding, F., Dokholyan, N.V., Buldyrev, S.V., *et al.* (2002). Molecular dynamics simulation of the SH3 domain aggregation suggests a generic amyloidogenesis mechanism. **J. Mol. Biol**. (324) 851-857.

57. Hamada, D., Segawa, S. and Goto, Y. (1996). Non-native alpha helical intermediate in the refolding of beta-lactoglobulin, a predominantly beta-sheet protein. **Nat. Struct. Biol.** (3) 868-873.

58. Selkoe, D.J. (1996). Amyloid beta-protein and the genetics of Alzheimer's disease. **J. Biol. Chem.** (271) 18295-18298.

59. Bradbury, J. (2003). Chaperones: keeping a close eye on protein folding. **The Lancet** (361) 1194-1195

60. Vainberg, *I.E.*, Lewis, S.A., Kellermann, J., *et al.* (1998). Prefoldin, a chaperone that delivers unfolded proteins to cytosolic chaperonin. **Cell** (93) 863-873.

61. Thulasiraman, V., Yang, C.F. and Frydman, J. (1999). In vivo newly translated polypeptides are sequestered in a protected folding environment. **EMBO J.** (18) 85-95.

62. Cohen, E.F. (1999). Protein misfolding and prion disease. **J. Mol. Biol.** (293) 313-320.

63. Helen, R. S. and Neil, A.R. (2002). The chaperonin folding machine. **Trends in Biochemical Sciences** (27) No. 12, 627-632.

64. Claudio, S. (2001). Protein misfolding and disease; protein refolding and therapy. **FEBS letters** (*498*) 204-207.

65. Claudio, S. and Gabriela, P.S. (2001). Prions: disease propagation and disease therapy by conformational transmission. **Trends in Molecular Medicine** (7) 109-114.

66. Soto, C., *et al.* (2000). Reversion of prion protein conformational changes by synthetic b-sheet breaker peptides. **Lancet** (355) 192-197.

Advances in Biochemistry and Biotechnology Volume 1 *Pages 107-121*

Edited by: **Dr. Chiranjib Chakraborty**

Published by: **DAYA PUBLISHING HOUSE**

5

Recombinant Therapeutic Proteins from Bacteria, Animal and Plant Cells

J.K. Deb

Department of Biochemical Engineering and Biotechnology, I.I.T., Delhi, New Delhi – 110 016

ABSTRACT

In the past two decades, the growth of biotechnology industries has witnessed great progress in the marketing of protein therapeutics which include cytokines, growth factors, hormones, blood clotting factors, enzymes, peptide drugs, antibodies and the list is growing rapidly. Spectacular advances in gene expression technology have made this progress possible. Beginning with the first recombinant human insulin produced by Eli Lilly in 1980s, majority of the recombinant proteins were produced by *Escherichia coli*. The trend has changed during the last decade. Due to the limitations of *E. coli* in folding protein, secretion of proteins in the medium and to carry out glycosylation, alternative gene expression systems have been developed. Gram positive, nonpathogenic bacteria hold promise as alternative bacterial expression systems in situations where glycosylation is not so vital for the biological activities of the therapeutic proteins. At present bacteria, yeast and animal cells constitute the most common expression systems for the production of such therapeutics. Increasing interest is now focussed on transgenic animals and plants because of their ability to produce complex proteins at low cost.

Introduction

Biopharmaceuticals have come to occupy centre stage globally in the area of therapy of both human and animal. The actual boost came towards the last 4 decades of the century just gone by when biomolecules with therapeutic potential such as insulin, antibodies, were used for treatment of diseases and ever since the demands were growing. The basic problem was the availability of these and other biomolecules since these were obtained by direct extraction of from the native sources in exceedingly low quantities. This approach was not practically viable. Besides, presence of contaminants and immunologic reactions could also pose problems in specific cases. Although synthetic and semisynthetic approaches were available at a practical level for small and medium sized molecules, these were not viable for protein therapeutics. Demand for the latter had to wait till important breakthroughs in recombinant DNA technology and the development and production of monoclonal antibodies could be exploited. In 1982, less than ten years after the above breakthrough, the first recombinant protein in the form of human insulin was made available in the market. From 1985 onwards, recombinant proteins have gained enormous importance as therapeutic agents. The total sales of therapeutic proteins was US$9000 in 1994 which is expected to hit 20 billion US dollars in 2000[1] **(TIPS 1996, 17, 450-456)**. During last ten years or so a large number of biopharmaceuticals have received license for medical use. Currently, dozens of biopharmaceuticals are in the market under various trade names[2].(*www.bio.org/aboutbio/drugs.html*)

The area of recombinant protein therapy is getting the limelight in recent times for obvious reasons and there are large number of reviews, series and books available on various therapeutic proteins that description of the entire gamut of such an area of activity in one review is formidable, if not impossible. Therefore, this review will concentrate mainly on various expression systems currently in use for the production of therapeutic proteins like cytokines, hormones, blood clotting factors, thrombolytic agents and vaccines that are either in use or at various stages of development. Recent developments in the alternative expression strategies will also be reviewed.

Most of the therapeutic proteins currently in the market are naturally occurring human or animal proteins such as insulin, growth hormone, tissue plasminogen activator, various growth factors etc. and their modified derivatives. Proteins of viruses mainly constitute recombinant vaccines. Therefore, it is evident that the success of the technology for the production of these molecules depends on the availability of an efficient expression system. As already mentioned, there are several advantages of production of recombinant proteins. It is now possible to produce drugs like G-CSF, erythropoietin which could not so far be produced by conventional methods. The drugs such as insulin or growth hormone which are available in limited amounts can now be produced in high yield. Also, the provision of a safe product free from pathogens can be better ensured by this approach.

Majority of the currently used therapeutic proteins are glycoproteins. However, a good number of such proteins do not require sugar moiety for their biological activity, *e.g.*, IL-2, GM-CSF, G-CSF. The expression system to be used for a given protein, therefore, shall have to be based on this consideration. Other factors that influence the decision are desirability of secretion, the stability and refoldability of the protein and the projected dose of protein per patient. The latter determines whether the cost of the drug becomes critical.

Bacterial Expression Systems

Escherichia coli

Among bacteria, *E. coli* occupies the position of prominence for sound scientific reasons. No other organism is better understood in terms of genetics, physiology or biochemistry than *E. coli.* It is this wealth of information that has led to the present high level of sophistication in *E. coli* recombinant gene expression. Overexpression of a protein is required to be approached systematically taking stock of major issues such as choice of a vector, optimization of expression by making sequence changes and experimenting with genetic background of the host, and finally optimization of growth conditions. It is likely that maximization of all aspects of expression is unattainable because eventually one step will become rate limiting. Thus optimization may be achieved through a combination of choices that can be changed or substituted as the situation demands.

A well designed prokaryotic expression system contains a set of optimally configured genetic elements that both transcriptional and translational aspects of protein production. One of the most important elements is the vector, which is normally a plasmid. High copy number vectors normally ensures high level of production but issues such as toxicity resulting from high gene dosage of certain genes may render such an approach counterproductive. In a situation like this, expression from a strong promoter may be tightly regulated by repression followed by induction. Excellent review has appeared recently on the fundamentals of *E. coli* expression vectors (Beneyx, 1999, Curr. Opin. Biot., 10, 411-421). Many strong promoters are now available in *E. coli*. They belong to different categories. Some are natural, others are hybrid, synthetic or mutagenized. By and large, all the systems are regulated by various inducers. Various forms of induction have been in use with *E. coli*. Common among these are *lacI*, *trp*, *pho* based on induction respectively by IPTG, tryptophan starvation and phosphate starvation. Alternatively, introduction of temperature sensitive mutant of *lacI* gene that encodes a thermosensitive *lac* repressor provides a convenient method to induce *lac* based promoters[3,4]. A disadvantage of this strategy is the upregulation of proteases following heat shock response. This problem is minimized by the use of host strains with a deficient *rpoH* (*htpR*) locus. Another regulatable promoter now being used is *araBAD* inducible by arabinose. Using this promoter, tissue plasminogen activator (tPA) has been produced in *E. coli*[4]. Of all these promoters the one that is inexpensive and subject to automatic induction (*i.e.*, requires no operator intervention) is *phoA* as it desirable in case of large scale production. Since induction requires phosphate starvation, it could limit the duration of protein synthesis. Expression yields can be improved from *phoA* promoter wothout loss of control when PhoS phosphate sensing protein was mutated to allow promoter induction at higher phosphate concentrations. Phosphate can then be fed continually (to avoid phosphate starvation) without repressing the promoter.

Another problem experienced with expression in *E. coli* has to do with its inability to form disulphide bonds in proteins with multidisulphide bonds. The formation of disulphide bonds in *E. coli* is a complex process involving several genes *viz.*, *DsbA, B, C and D* of which *DsbA* is a oxidoreductase and *DsbC* is a disulphide isomerase. In spite of this it has been observed that that several proteins, especially of eukryotic origin, such as tPA, various growth factors, protease inhibitors which contain substantial number of disulphide bonds can be folded properly in this organism and one has to rely on their expression in eukaryotic system. However, coexpression of *DsbA* and *DsbC* has made it possible to obtain recombinant tPA having as high as 17 disulphide bonds in physiologically active form in *E. coli*. Overproduction of proteins very often leads to the misfolding of proteins in the cytoplasm leading to the formation of insoluble aggregates called inclusion bodies. Although this simplifies purification of proteins from the cytoplasmic milieu, the recovery of native protein requires further manipulation. Proper folding of protein from the inclusion bodies requires control of aggregation. Several approaches have been employed to achieve this. Recent review has dealt with this topic[5]. Tran-Moseman *et al.*, obtained renaturation of human macrophage colony stimulating factor (hM-CSF) from the reduced, denatured monomers using thiol mixture glutathione and oxidized glutathione and a low concentration of denaturing agent (urea or guanidinium chloride). They obtained higher renaturation yield using purified reduced, denatured protein. A maximum of 95% refolding was obtained under this condition[6].

Several chaperones have been implicated in the folding process. In vivo, these chaperones participate by transient interaction with folding intermediates during the folding thereby promoting proper isomerization and cellular targeting. The complex process involves participation of ATP dependent DnaK-DnaJ-GroE and GroEL-GroES systems of chaperones[8]. Finding the optimal refolding conditions is still relatively empirical, with the best results obtained from evaluating a variety of conditions affecting solubility and disulphide bond formation and isomerization. The use of either soluble or immobolized folding aids can sometimes be effective. Altimarano *et al.*, .[9] could improve the folding yield of scorpion toxin Cn5 from less than 5% to more than 85% by immobilization of a minichaperone, prolyl isomerase and the periplasmic oxidoreductase DsbA. Interestingly, Berges *et al.*, have reported export of GM-CSF and Interlukin 13 to periplasm by fusing the coding sequences of these cytokines with two synthetic signal sequences and observed that both the total amount of protein and the fraction of soluble protein in the periplasm depended on the signal peptide-protein combinantion. Overexpression of SecB or DnaK plus DnaJ resulted in marked increase in the amount of protein in the periplasm, but it also depended on the signal peptide-cytokine-chaperone combination[10].Although periplasm is where the proteins can better be folded in contrast to the cytoplasm of *E. coli*, it has not evolved to support folding of large, complex proteins with many disulphide bonds. However, modification of folding environment can sometimes give results as mentioned above for tPA[11]. In the case of IGF-I, with three disulphide bonds, periplasmic expression of DsbA or DsbC actually decreased the yield of folded product, although it significantly increased its accumulation in the inclusion bodies[12].

Inability of *E. coli* to secrete proteins has been a limitation from downstream processing viewpoint. Attempts to selectively release recombinant proteins from this organism into the surrounding medium have not met with much success. Recently, it was shown that overexpression of third topological domain of transmembrane protein TolA (TolAIII) in the periplasm makes the cell leaky by disrupting the integrity of the outer membrane and releasing periplasmic proteins into the medium[13]. Unfortunately, the bacteria became fragile and suffered loss in viability.

For high level production and purification of proteins in *E. coli*, several sophisticated protein-fusion systems have been developed. Fusion partners offer several advantages, such as prevention of inclusion body formation, improved folding characteristics, limited proteolysis and easy protein purification schemes. The most common fusion proteins used are glutathione-S-transferase (GST), staphylococcal protein A, streptococcal protein G, maltose binding protein (MBP). Sometimes, affinity tag, such as His_6, FLAG etc. are also used. All these permit purification of fused proteins using specific ligand.

Gram Positive Bacteria

Bacteria with Low GC Content

Many gram positive bacteria are useful for their medical, food and biotechnological potential. Another important property of these organisms is their secretory power. Industrially important, nonpathogenic bacteria, such as *Bacillus, Streptomyces,* are well studied as far as their genetic manipulation is concerned. Efficient expression vectors are available for these bacteria. Gram positive bacteria such as these belong to two distinct groups based on the GC content of their DNA. While *Bacillus* and related genera have low GC content, *Streptomyces* has high GC content. Among the low

GC content gram positive bacteria, those belonging to the genera *Lactococcus, Lactobacillus,* some nonpathogenic species of *Streptococcus* and *Staphylococcus* are useful for the production of therapeutic proteins and vaccines. Among high GC content gram positive bacteria, the nonpathogenic *Corynebacteria* are also well researched as far as the genetic manipulation is concerned. Bacteria such as *Bacillus subtilis, Bacillus brevis, Bacillus licheniformis, Bacillus megaterium, Corynebacterium glutamicum, Streptomyces, Lactobacillus and Lactococcus* are safe orgainisms, and hence, suitable for the production of therapeutic proteins. Although *B. subtilis* is the most important member of this group, its main drawback is high level secretion of various proteases. Several mutants lacking in two of the major proteases and some lacking other minor proteases have been isolated to get around the above limitation. Using a *B. subtilis* mutant, WB600, deficient in 6 extracellular proteases, Wong *et al.*,[14] secreted streptokinase, the popular thrombolytic drug synthesized by *Streptococcus equisimilis,* with the help of inducible SacB promoter and its signal sequence fused to the streptokinase gene and also with its own promoter and signal sequence. The mature protein was found to have the same N-terminal sequence as that of the native streptokinase from *S. equisimilis. B. subtilis* was also used to produce staphylokinase, a 136-amino acid long thrombolytic protein secreted by *Staphylococcus aureus*. Dramatic improvement in the production of this protein was observed when the mutant in which the gene for subtilisin was disrupted was used. This strategy also improved the stability of the extracellular staphylokinase[15]. An alternative bacillus species having negligible protease activity is *Bacillus brevis*. Significant importance has been attached to this strain which also produces large amount of S layer proteins. Based on these proteins, expression-secretion vectors have been constructed that allowed for the extracellular production of large amount (more than 1g/l) of functional Epidermal Growth Factor (EGF)[16]. In another study, erythropoietin (EPO) and its extracellular receptor were also produced, albeit less efficiently[17]. Bacteria belonging to the genuses *Streptococcus, Staphylococcus, Lactococcus, Lactobacillus* have mostly been used in vaccine development. Among the organisms belonging to the first two classes are those which are food grade bacteria, such as *Staphylococcus carnosus, Staphylococcus xylosus, Streptococcus thermophilus* or *Streptococcus gordonii,* the latter being a commensal bacterium of human oral cavity. All these strains were used to display recombinant protein on the surface using the anchor domain of staphylococcal protein A. Samuelson *et al.*, reported the display of 80 amino acid peptide from a malaria blood stage antigen on the surface of *S. carnosus* and verified the presence on the surface by immunogold staining and immunofluorescence[18]. The aim was to use nonpathogenic staphylococci with a long tradition as novel vaccine carriers. Similarly, strains of both *S. carnosus* and. *xylosus* appeared to be excellent hosts for surface production of functional anti-human IgE single-chain Fv antibodies and mouse IgG1(κ)[19]. Pozzi *et al.*, used *S. gordonii* to display immunodominant epitope of HIV-1 gp120 and it was found to be recognized by human T helper cells[20]. Lactic acid bacteria such as *Lactobacillus* and *Lactococcus* assume great importance in oral vaccine development due to their presence as normal microflora of gut. Among lactobacilli *Lactobacillus plantarum* has received considerable attention due to its capacity to secrete protein to a reasonable degree. With this bacterium, α-amylase and levanase have been secreted using homologous expression-secretion signals both located on the chromosome and on plasmids[21]. In recent times, several constitutive and regulated expression systems have been developed in *Lactococcus lactis*, widely used as a starter in cheese production. A T7 promoter-T7 RNA polymerase based expression system of this bacterium has been used to express murine IL-2 extracellularly[22]. Both lactobacilli and lactococci secretion systems based on S layer protein of *lactobacillus brevis* have been developed[23]. Great deal of studies on the adjuvanticity and immunogenicity of these two classes of bacteria have been carried out. *Lactobacillus casei* var. Shirota and *L. plantarum* ATCC 8014 showed strong adjuvant effects, equivalent to that of common water-in-oil emulsion adjuvant, Specol[24]. Administartion of lactobacilli and lactococci by parenteral route induce the formation of antibodies of

IgG class in experimental animals. Both types of bacteria have been used to express antigens intracellularly, extracellularly or on the surface. Mice immunized with live recombinant bacteria by parenteral, oral or nasal routes and significant immune response was observed following mucosal administartion when the priming reaction was followed by a booster immunization. Parenteral immunization resulted in antibody production following priming immunizations only[25]. High level of expression of tetanus toxin fragment C and protection against lethal challenge was observed with recombinant *L. lactis*[26,27].

Gram Positive Bacteria with High GC Content

Major bacteria of this class are members of actinomycete group. The importance of these bacteria are either due to their pathogenicity, as in the case of *Mycobacteria, Nocardia* or due to their therapeutic or industrial uses, as in the case of *Streptomyces*, certain species of *Nocardia, Coryenbacteria*, except several species of the latter, *e.g., Corynebacterium diphtheriae*, which are pathogenic. Of these organisms, *Streptomyces* has been used for the production of several therapeutic proteins in recent times. The advantage of using this organism is its well understood molecular genetic systems. Efficient cloning and expression vectors, including inducible vectors, are available for these bacteria. The best studied of the *Streptomyces* species is *Streptomyces coelicolor* A(3). The other common species is *Streptomyces lividans. Streptomyces* species have been useful for expressing other actinomycete genes. For example, inducible expression of histidine-tagged *Nocardia* proteins has been made possible in *S. lividans* using *tipA* promoter, which can be induced by thiostreptone in *streptomyces*[28]. Although *S. coelicolor* A3(2) is the most extensively characterized species, *S. lividans* 66 is almost exclusively the species of choice as a host for heterologous protein production. The reason for this is that *S. coelicolor* A3(2) possesses a very powerful restriction system which *S. lividans* lacks. In addition, *S. lividans* is also unlike most other streptomyces in that it generally displays very low level of endogenous extracellular proteolytic activity and at the same time retains the ability to produce a large quantities of extracellular protein.

Of all the actinomycetes, perhaps *Streptomyces* is the organism in which maximum number of therapeutic proteins have been expressed. Thus, Koller *et al.*,[29] have produced proinsulin with an yield of.300 mg/L. Besides, TNF α[30], GMCSF[32] and IL 1β[33] were produced in quantities ranging from 10–100 mg/L, except in case of IL1β, where the yield was 1-5 mg/L.

Another member of actinomycetes which has the potential for such exploitation is food grade corynebacteria such as *Corynebacterium glutamicum*. This strain and other related species have no detectable extracellualr proteases or toxins[34]. Another advantage of this system is similarities of various regulatory processes between corynebacteria and *E. coli* and efficient functioning of common *E. coli* plasmids like *tac, lacUV, trp* in corynebacteria[35]. Various expression and secretion vectors are now available in this system[36-38]. For easy purification of recombinant proteins, gene fusion vectors with streptavidin and Glutathione-S-transferase (GST) have also been constructed[39,40]. Using GST-fusion vector, pBKGEXm2, streptokinase has been produced in *Corynebacterium acetoacidophilum*[40]. One important finding of this work is the presence of bulk of the streptokinase in the medium, indicating secretory nature of this system. Recently, these authors have developed a corynebacterial secretion vector which efficiently secretes heterologous proteins into the culture medium (unpublished observations).

Yeast Expression Systems

Saccharomyces, Hansenula, Kluiveromyces

Yeasts are simplest of eukaryotic organisms capable of performing eukaryotic processing steps of proteins and thus constitute attractive systems for heterologous gene expression. Of these *Saccharomyces cerevisae* is a GRAS level organism and most widely studied yeast, as regards its basic genetics and biotechnology. Since excellent reviews are available on the expression of heterologous genes by these orgainism, the present discussion will be limited to the brief mention of the promoters and vectors used for the purpose. Most of the vectors used for cloning and expression are *E. coli-S. cerevisae* shuttle vectors. Various versions of vectors are available. Depending upon their mode of replication they are classified into three major types, *viz.*, YIp, YEp and YRp. The former does not replicate independently and gets integrated into the chromosome. Consequently, these plasmids have low copy number. These gene dosage limitation has been resolved either by employing yeast transposon to create multiple insertions or by targeting the cloned gene to ribosomal gene cluster[41]. The vector of Yep type are usually present in high copy number (30 or so) and replicate independently of the yeast chromosome due to the presence of *ARS* sequence and the *STB* locus derived from yeast relicating 2μ plasmid. One disadvantage associated with instability of these vectors have been overcome by introducing *srb1-1* mutation that makes the host strain osmotically sensitive, the wild type gene *SRB1* incorporated in the vector complements the deficiency of the host strain[42]. The expression of heterologous genes have been achieved using either a constitutive or an inducible promoter. Some of these promoters belong to genes of highly expressed glycolytic enzymes, such as *GAPDH, PGK*. Other important *S. cerevisiae* promoters are *ADH1, GAL1/GAL10, PHO5, MF*α*1*. Several important therapeutic proteins such as 1 antitrypsin, hirudin, HbsAg, HuIFN α, HuEGF, HuSOD, human serum albumin (HSA) have been produced in *S. cerevisiae*. Although initially *E. coli* was tried for production of HbsAg, it failed to make complex particle of antigen necessary for its immunogenicity. Until recently, *S.cerevisiae* occupied the exalted position of being the sole organism for the industrial production of HbsAg and insulin globally. Recently, *Pichia pastoris* is being used for the production of HbsAg.

Pitchia pastoris

P. pastoris has become another popular yeast system for the production of recombinant proteins and nearly 300 foreign proteins have been produced since 1984 using this system. This yeast, besides *Hansuela, Candida and Torulopsis*, share a general methanol utilization pathway. In methanol grown cells, the key enzymes of this pathway constitute 30% of total cellular proteins. Two compelling factors for the popularity of *P. pastoris* are: i) the availability of a strong alcohol oxidase I (*AOX1*) promoter, the strongest and most regulated promoter known, and ii) the ability to integrate expression plasmids at specific sites in its genome. The ability to grow on methanol has given an added advantage to this organism, because mineral medium consisting of methanol, biotin, salts and trace elements are economical and well defined, ideal for large scale production in fermenters. It can produce protein at very high level and carry out higher eukaryotic modification like glycosylation, disulphide formation and proteolytic processing. For this reason, many proteins which could not be produced in active forms in bacteria, *S. cerevisiae* and baculovirus, have been produced in this organism.

The *AOX1* promoter is tightly repressed by glucose and most other carbon sources. Therefore, with this promoter the expression remains repressed by growing cells with any of the above carbon source, especially glucose or glycerol, to a high cell density followed by derepression by adding methanol as sole carbon source whereupon the induction level exceeds 1000 fold. Several therapeutic human proteins, such as insulin, HSA, chitinase, antithrombin II, type I collagen, malaria vaccine antigen, bile salt stimulated lipase, antibody cahin fragment C of botulinum antitoxin etc. have been produced in fermenters by *P. pastoris*. All the above proteins, except the last mentioned one, were secretory products and were produced at levels of several milligrams to gram per litre. The yield of insulin was 1.5 g/L[43].

There is a limitation with *AOX1* promoter because of its induction with methanol, a toxic and hazardous substance. Several other promoters, such as constitutive glyceraldehydes-3- phosphate dehydrogenase (*GAP*) and inducible glutathione dependent formaldehyde dehydrogenase, *FLD1*, promoters have been used. Constitutive and high level expression with these promoters are also toxic to the cells. So, a moderately expressed promoter such as *PEX8* promoter ensures modest induction when cells are shifted from glucose to methanol medium. Even in this case also induction by methanol becomes a necessity.

Transgenic Plants as Expression System

The use of plants as production system for foreign proteins has gained wide spread support due to several reasons. Producing therapeutic proteins in plants has many economic and qualitative benefits, including reduced health risks due to pathogen contamination, comparatively high yield, production in seeds or other storage organs. Overall, transgenic plants are potentially cheap sources of recombinant proteins and it is estimated that the cost of production in plants could be 10- to 50- fold cheaper that in recombinant *E. coli*[44].

Besides economic advantages, there are qualitative benefits in producing recombinant proteins in transgeneic plants. Animal expression systems are expensive and sensitive to environmental changes when cultured on an industrial scale. Animal cells are very sheer sensitive and so require special design of reactors. Careful control of culture conditions is required to ensure product purity[45]. Although microbial systems are robust, they are not ideal for the synthesis of many mammalian proteins because of differences in processing and posttranslational modifications. Such differences are not altogether non-existent between plants and animals. Nevertheless, these issues deserved to be sorted out to make such an enterprise a success. For example, high mannose type N-glycans found in plant derived antibody have β (1,2)-xylose and α (1,3)-fucose residues linked to the core Man3GlcNAc2. Being absent in animals, such linkages are potentially immunogenic and may have limited scope as parenterals or for topical or oral use, specially in patient with severe food allergies. One major development in this direction has been the stable cloning of human β 1,4-galactosyl transferase in tobacco plants[46]. This enzyme, missing in plant, is responsible for conversion of typical plant glycans into human glycans. A cross of a tobacco plant expressing the above enzyme with a plant expressing heavy- and light- chains of mouse antibody resulted in the expression of a an antibody in which galactosyl glycan is 30%, a level that is about as abundant as the same antibody produced by hybridoma cells.

Another major problem in the expression of heterologous protein in plants has to do with the silencing of transgene. Post-transcriptional gene silencing (PTGS). It is often associated with methylation of transcribed region of the silenced gene and with the accumulation of small RNA molecules homologous to the silenced gene. The most efficient way of avoiding PTGS was to design carefully transgene avoiding repeat sequences and analyzing the transformants at the molecular level. Another strategy is to introduce a helper-component proteinase (HC-Pro) of plant poty virus into plant, which results in the loss PTGS. HC-Pro brings this about by targeting a PTGS maintenance component at a point that affects accumulation of small RNAs and methylation of genomic DNA[47]. Notable progress has been achieved in recent years in reducing gene silencing problem and position effects of nuclear transformation by introducing a foreign gene between the functional genes of chloroplast using homologous recombination. High level expression of oligomeric cholera toxin B (4-5% of total soluble protein) and human somatotropin (7% of total soluble protein) has been achieved by such a strategy[48].

Although plant is projected as a low cost production system for protein expression, downstream processing of plant-derived product is potentially costly process. In a recent cost analysis of β-glucuronidase production from transgenic corn seed, milling, protein extraction and purification accounted for 94% of the production cost. One way to reduce this to direct protein synthesis to seed endosperm from where proteins could be extracted. Thus, avidin was extracted from transgenic maize seed endosperm at a cost 10 fold cheaper than that conventionally purified from eggs[49]. Different strategies are used to direct protein accumulation in different parts of the plant. For example, hirudin, an anticagulant used for treating thrombosis, has been produced as oleosin-hirudin fusion protein which allowed for extraction of the complex from oil bodies of recombinant oilseed rape[50]. Using various promoters transgenic proteins have been deposited in specific parts of the plant. Thus, using promoters of genes with aleurone specific expression during germination such as α-amylase and signal export peptide recombinant β-glucanase could be exported to the endosperm. On the other hand, using promoter of structural genes for storage proteins such as D hordein (*Hor3-1*) promoter, the same protein was produced in maturing grain in barley[51].

In recent times, *in vitro* plant cell culture technology has become an important tool for the expression of foreign protein. This may be the most suitable system for medium level production of high price, high purity speciality molecules, such as therapeutic proteins. However, plant cell culture is unlikely to be competitive for the majority of proteins required in large quantities that is otherwise deliverable by agricultural means. It may be useful if the product is accumulated at levels at least one order of magnitude higher than those achieved by agricultural systems. This would reduce the cost associated with product recovery to offset the substantially higher cost of biomass production. Another advantage with plant cell culture is that the time involved for production is much shorter than the growth cycle of whole plants. Therefore, compared to months needed for agriculture, proteins would be manufacture within days or weeks. From the safety and containment points of view, plant cell culture is advantageous as it is amenable to growth in bioreactors under controlled conditions.In addition, when the protein of interest is secreted into the medium of cell cultures, product recovery and purification could be greatly simplified, while eliminating the need to destroy the cells.

Mammalian Expression Systems

During past nearly a decade there has been considerable progress in the development of mammalian expression systems for high level recombinant protein production. The expression can be stable or transient. In stable expression technology, which is based on chromosome-integrated plasmid sequences, good quantities of complex proteins from mammalian cells can be obtained[52]. On the other hand, milligram to gram quantities of protein can be obtained by transient expression strategy. The transient transfer of DNA to mammalian cells serves an important role in the manufacture of products involved in somatic gene therapy. Such a system has several key features like i) simplicity in the construction of expression vectors, ii) short duration for the generation of product (days), iii) intrinsic genetic stability and consistencey due to short period elapsed between generation of vector and product recovery, iv) applicability to a wide range of host cell lines.

Cell types that dominate the field of recombinant mammalian system are chinese hamster ovarian (CHO), COS, baby hamster kidney (BHK) and NSO murine myeloma cell lines. Refinements of vector construction, use of strong and regualatable promoters, choice of selectable markers have made the establishment of recombinant cell lines with high specific productivities. Normally, the vectors are either plasmid based or virus based. Strong promoters commonly used are derived from SV40 or CMV virus[53,54]. One nonviral promoter, the elongation factor (EF-1) promoter, has attracted attention because it appears to be as strong or stronger than some viral promoters[55]. Use of elements that provide extended maintenance of plasmids in cells have now been possible in transient expression of genes. These vectors are used in combination with cell hosts that contain elements for the promotion of episomal plasmid replication, such as large T antigen from SV40. The plasmid would then carry the origin of replication from SV40. Also, identification of 'hot spot of integration' within the chromosome resulting in stable transformation and the enhancement of biomass accumulation has increased productivity of recombinant mammalian cells, *e.g.*, an yield of 1-2 g/L of unpurified antibody has been with NS0 myeloma cells[56]

Alphavirus, adenovirus and vaccinia virus are widely used for construction of viral vectors. Sindbis virus and Semliki Forest virus (SFV) are the most studied alphaviruses used for heterologous expression[57]. The broad host range of SFV has particularly made it attractive. In order to use as a safe vector, two plasmids are made from SFV, one for cloning, pSFV1, which carries the replicase genes and the other, the helper plasmid, that codes for the structural proteins of the virus. To improve the safety condition, the system was modified to allow the production of non-infectious SFV particles, which can be subsequently activated by enzymatic cleavage of the spike protein. The extremely efficient SFV 26S promoter used and high copies of RNA (200,000) per infected cell lead to high expression levels. Recombinant proteins are produced with post-translational modifications conferred by the specific processing capacities of the host cell selected[58, 59]. This system has also been scaled up[60]. The extensive knowledge accumulated over the years on adenovirus transcription regulation allowed the engineering of adenovirus vectors modified for heterologous expression. For this purpose, the early regions of E1 and E3 were deleted to render the virus replication incompetent and this function must be provided in trans by the host cell. An expression cassette set in the deleted E1 region and containing additional (or ectopic) major late promoter, such as CMV was used to clone the foreign gene. A series of vectors, pAdBM5, was constructed with enhancers added to the ectopic major late promoter region[61]. The ability of the virus to be propagated in suspension culture is of considerable advantage for large scale work[62]. One of the problems with adebnoviral vectors is the elicitation of humoral and cellular immune response directed against the viral gene products, albeit at a low level, even in the absence of E1A gene product. Newer vectors which are less immunogenic gave been developed, which are defiant or defective in the E2, 3 and E4 gene regions. These vectors can be propagated in trans complementing cell lines. Some vectors, called "gutless vectors" have been developed in which only the vector inverted terminal repeats (ITRs) and viral packaging signal are present. Theoretically it can accommodate upto 35 kb of insert but requires all the viral gene products in trans that are needed for virus assembly. Vaccinia virus genome is double stranded DNA virus of nearly 200 kb and replicates in the cytoplasm of the host cell. Infection with this virus often leads to 5000viral particles per cell, which leads to high level recombinant protein expression. The recombinant gene is placed under the control of viral promoter. Normally, late promoters are preferred or compound early/late promoter may be chosen to minimize the cytopathic effect. Advantages with vaccinia virus are that integration into the chromosome can be done by homologous recombination, direct cloning and wide host range. A disadvantage with this system is that high safety level is required. Vaccinia system has been successfully used at very large (1000 liter) to produce many different proteins[63].

One of the notable areas of advances in the application of genetic engineering approaches to rationally modify specific features of mammalian host to improve their utility in recombainant protein expression. Foe example, work during the past decade has shown how various reactions in the glycosylation pathway can be influenced by cell culture factors, host cell selection and protein specific features, leading to the production of molecules with variable or suboptimal clearance or bioactivity properties[64,65]. Studies have shown that specific manipulation of oligosaccharide structures on a recombinant protein by overexpression of appropriate glycosyl transferases can enhance glycan quality. For example, overexpression of a galactosyltransferaseand a silayltransferase in CHO cells led to corresponding increases in the galactose and sialic acid content of expressed protein[66].

Another important factor has been to improve the efficiency of central carbon metabolism and to reduce lactate accumulation. In one approach, partial disruption of lactate dehydrgenase A gene led to reduced lactate accumulation and improved cell culture performance in hybridoma production[67]. Perhaps the most interesting of the targets for metabolic control have been that of cell growth and the limit of apoptotic cell death. The ability of chemical inhibitors of apoptosis to extend culture viability has opened up the growing area work directed towards genetic engineering of cell death pathways. Overexpression of Bcl-2 and Bcl_{xL} has been shown to inhibit apoptotic death in several mammalian culture systems[68]. A recent report also showed increased volumetric antibody expression in butyrate treated CHO cultures overexpressing Bcl-2[69].

Conclusion

In recent years, high demand for therapeutically active protein molecules has spawned intense research in the development diverse expression systems for the production of these potent biomolecules. Several organisms spanning bacterial, yeast, fungal, plant and animal cells have been used as systems for the purpose depending upon the regulatory need and bioactivity of the product. Although *E. coli* commanded highest applicability and capacity utilization is currently 90%, not all therapeutic proteins can be produced in *E. coli* because the bacteria are unable to modify proteins after they are produced, which can cause misfolded and inactive proteins. For these reasons, monoclonal antibodies, fusion proteins and other proteins that must be modified must be produced ideally in mammalian cells, such as CHO, NS0, BHK etc. Being more cost intensive, other eukaryotic systems, such as yeast, fungi are sometimes exploited as substitute for mammalian system. Even such a system may not be ideal where glycosylation akin to mammalian system becomes a necessity.

The role of plant tissue cultureas a means for commercial protein production is not clear at present, but is likely to depend upon the ability to express high levels of proteins in vitro by manipulating culture conditions. Extracellular localization and the stability of secreted protein in the medium of plant cultures may be important keys to economic feasibility, allowing for easier product recovery without destroying biomass. Because the controlled environment in plant cultures offers advantages for good manufacturing practice (GMP) and regulatory compliance, a potential niche for reactor-scale plant tissue culture is the rapid production of low-to-medium volume therapeutic proteins. Technology for modifying the glycans of plant glycoproteins may be required before systems can be fully exploited for the manufacture of injectable drugs and vaccines.

Mammalian derived cell production of biopharmaceuticals has a good regulatory track record, and allows for the modification of proteins with reduced immunogenicity. Compared to production by *E. coli* or other bacterial system, however, the construction of genetically engineered cells is more time consuming, production is less efficient and more expensive, and purification is more complicated.

REFERENCES

1. Peter Buckel (1996) Recombinant proteins for therapy Tr. in Pharmacology, 17, 450-456.
2. www.bio.org/aboutbio/drugs(html <http://www.bio.org/aboutbio/drugs.html>)
3. Gerhard Hannig and Savvas C. Makrides (1998) Strategies for optimizing heterologous protein expression in Escherichia coli Tr. in Biotechnol. 54, 16

4. Hasan, H. and Szybalski. W. (1995) Construction of *lacI*ts and *lacI*^{q}ts expression plasmids and evaluation of the thermosensitive *lac* repressor Gene 163, 35-40

5. Poole, E.S., Brown, R. E. and Buckel, P. (1995) The identity of the base following the stop codon determines the efficiency of in vivo translational termination in Escherichia coli EMBO J. 14, 151-158

6. Qui J, Swartz, J.S. and Georgiou, G. (1998) Expression of Active Human Tissue-Type Plasminogen Activator in *Escherichia coli* Appl. Env. Microbiol. 64, 4891-4896.

7. Lilie H, Schwarz, E and Rudolph. R. (1998) Advances in refolding of proteins produced in *E. coli* Curr. Opin. Biotech., 9, 497

8. AnhDao Tran-Moseman, Neil Schauer and Eliana De Bernardez Clark (1999)

9. Renaturation of*Escherichia coli*-Derived Recombinant Human Macrophage Colony-Stimulating Factor Protein Express Purif, 16, 181-189

10. Richardson. A, Landry. S.J. and Georgopoulos. C (1998)The ins and outs of a molecular chaperone machine Tr. in Biochem. Sc., 23, 138-143.

11. Myriam M. Altamirano, Consuelo García, Lourival D. Possani, Alan R. Fersht (1999) Nat. Biotechnol. 17, 187-191

12. H Berges, E Joseph-Liauzun, and O Fayet (1996) Combined effects of the signal sequence and the major chaperone proteins on the export of human cytokines in Escherichia coli Appl. Env. Microbiol., 62, 55-60

13. John C. Joly, W. S. Leung, and James R. Swartz (1998) Overexpression of *Escherichia coli* oxidoreductases increases recombinant insulin-like growth factor-I accumulation Proc Nat. Acad. Sci., USA 95, 2773-2777

14. Wan E.w. and Baneyx. F. (1998) TolAIII Co-overexpression Facilitates the Recovery of Periplasmic Recombinant Proteins into the Growth Medium of *Escherichia coli* 14, 13-22

15. SL Wong, R Ye, and S Nathoo (1994) Engineering and production of streptokinase in a Bacillus subtilis expression-secretion system Appl. Env. Microbiol. 60, 517-523

16. Hee Bae, Si Myung Byun, and Jae Hoon Chung (2000) Enhancement of Secretion and Extracellular Stability of Staphylokinase in *Bacillus subtilis* by *wprA* Gene Disruption Appl. Env. Microbiol. 66, 476-480

17. Ebisu. S., Takagi. H., Kadowaki. K., Yamagata. H. and Udaka. S. (1996) The efficient production of epidermal growth factor by *Bacillus* brevis ANN NY Acad Sci, 782, 115-122.

18. Nagao. M., Inoue. K., Moon. S.K., Masuda. S., Takagi. K. and Udaka. S. (1997) Secretory production of erythreopoeietin receptor by *Bacillu brevis* Biosci Biotechnol Biochem, 61, 670-674.

19. P Samuelson, M Hansson, N Ahlborg, C Andreoni, F Gotz, T Bachi, TN Nguyen, H Binz, M Uhlen, and S Stahl (1995) Cell surface display of recombinant proteins on Staphylococcus carnosus J. Bacteriol. 177, 1470-1476

20. E Gunneriusson, P Samuelson, M Uhlen, PA Nygren, and S Stahl (1996) Surface display of a functional single-chain Fv antibody on staphylococci J. Bacteriol. 178, 1341

21. Gianni Pozzi, Marco R. Oggioni, Riccardo Manganelli, Donata Medaglini, Vincent A. Fischetti, Daniela Fenoglio, Maria T. Valle, Annalisa Kunkl and Fabrizio Manca (1994) Human T-helper

cell recognition of an immunodominant epitope of HIV-1 gp120 expressed on the surface of *Streptococcus gordonii* Vaccine 1994, 12, 1071-76

22. P Hols, T Ferain, D Garmyn, N Bernard, and J Delcour (1994) Use of homologous expression-secretion signals and vector-free stable chromosomal integration in engineering of Lactobacillus plantarum for alpha-amylase and levanase expression Appl. Env. Micro. 60, 1401

23. L Steidler, JM Wells, A Raeymaekers, J Vandekerckhove, W Fiers, and E Remaut (1995) Secretion of biologically active murine interleukin-2 by *Lactococcus lactis* subsp. Lactis Appl. Env. Micro, 1995, 61, 1627

24. Kirsi Savijoki, Minna Kahala and Airi Palva (1997) High level heterologous protein production in *Lactococcus* and *Lactobacillus* using a new secretion system based on the *Lactobacillus brevis* S-layer signals Gene 186, 255-262

25. Boersma.W.J.a., Bogaerts. W.J.A., Bianchi. A.T.J., and Classen. E. Adjuvant properties of stable water-in-oil emulsions:Evaluation of the experience with Specol. (1992) Res Immunol. 143, 503

26. Peter H. Pouwels, Rob J. Leer, Michael Shaw, Marie-Joan Heijne den Bak-Glashouwer, Frans D. Tielen, Egbert Smit, Beatriz Martinez, Jan Jore and Patricia L. Conway (1998) Lactic acid bacteria as antigen delivery vehicles for oral immunization purposes Int J Food Microb. 41, 155

27. Wells, J.M., Wilson, P.W., Norton. P.M., Gasson. M.J., and LePage. R.W.F.(1993) *Lactococcus* lactis: High-level expression of tetanus toxin fragment C and protection against lethal challenge Mol. Microbiol 8, 1155-1162.

28. Karen Robinson, Lisa M. Chamberlain, Karin M. Schofield, Jeremy M. Wells, and Richard W.F. Le Page (1997) Oral vaccination of mice against tetanus with recombinant Lactococcus lactis Nat. Biot. 15, 653-658.

29. Enguita. F.J., de la Fuente., J. L., Martín J. F and Liras P. (1996) An inducible expression system of histidine-tagged proteins in *Streptomyces lividans* for one-step purification by Ni^{2+} affinity chromatography FEMS Micro. Lett, 137, 135-140

30. JE Strickler, TR Berka, J Gorniak, J Fornwald, R Keys, JJ Rowland, M Rosenberg, and DP Taylor (1992) Two novel Streptomyces protein protease inhibitors. Purification, activity, cloning, and expression. J. Biol. Chem. 267, 3236-324.

31. Koller. K-P, Reiss. G., Klaus. S., Uhlmann. E., and Wallmeier (1989) Bio/technology 7, 1055-1059

32. Chang S-Y and Chang S 1988, Biology of Actinomycetes '88- Proc. of the 7th Intl. Symopsium on the biology of Actinomycetes (Okami, S-Y., Beppu, T and Ogawara, H.eds) pp 103-107, Japan Sc. Soc. Press

33. Malek L.T. In Pharm. Biotech. Intl., 1995 (Barnacal, P.A. ed), Sterling Publ., London, pp 55-57

34. Malek L.T., *et al.* (1990) Secretion of granulocyte macrophage colony stimulating factor (GM-CSF)in Streptomyces lividans. J. Cell. Biochem. Suppl. 14A, 127.

35. Lichenstein, H *et al.* (1988) J Bacteriol., 170, 3924-29

36. Billman-Jacobe (1996), Expression in bacteria other than *Escherichia coli* Curr. Opin. Biotechnol. 7, 500-504.

37. Eikmanns, B.J. *et al.* (1991) A family of *Corynebacterium glutamicum / Escherichia coli* shuttle vectors for cloning, controlled gene expression, and promoter probing. *Gene* 102, 93-98.

38. Patek, M. *et al.* (1996) Promoters from *Corynebacterium glutamicum*: cloning, molecular analysis and search for a consensus motif. *Microbiology* 142, 1297-1309.

39. Cadenas *et al.* (1996) Construction of new cloning vectors for *Brevibacterium lactofermentum* FEMS Microbiol. Lett. 137, 63-68.

40. Jakoby, M. *et al.* (1999) Construction and application of new *Corynebacterium glutamicum* vectors. *Biotechnol. Techniques* 13, 437-441.

41. Srivastava, P. and Deb, J.K. (2002) Construction of fusion vectors of corynebacteria: expression of glutathione-S-transferase fusion protein in *Corynebacterium acetoacidophilum* ATCC 21476. *FEMS Microbiol. Lett.* 212, 209-216.

42. Lopes T. S. *et al.*, 1989, High copy number integration into the ribosomal DNA of *Saccharomyces cerev.*, Gene 79, 199-206

43. Rech, S.B., Stateva, L.I.,and Oliver. S.G. (1992) Complementation of *Saccharomyces cerevisiae srb1* mutation Curr. Genet., 21, 339-344.

44. Cereghino, GPL *et al.*, , Production of recombinant proteins in fermenter cultures of the yeast *Pichia pastoris* (2002) Current Opinion in Biotechnology 13, 329-332

45. Kusnadi, A., Nikolov, Z.L. and Howard, J.A. (1997) Production of Recombinant Proteins in Transgenic Plants: Practical Considerations. Biotech. Bioengg., 56, 473-484

46. Giddings, G., Allison, G., Brooks, D and Carter, A. (2000) Transgenic plants as factories for biopharmaceuticals Nature, Biotechnology, 18, 1151-1155.

47. Bakker, H., Bardor, M., Molthoff, J.W., Gomord, V.V., Elbers II, Stevens, L.H., Jordi, W, Lommen, A., Faye, L, Lerouge, P., Bosch, D. (2001) Galactose-extended glycans of antibodies produced by transgenic plants Proc. Natl. Acad. Sci., USA. 98, 2899-2904.

48. Llave, C., Kasschau, K.D., Carrington, J.C. (2000) Virus-encoded suppressor of posttranscriptional gene silencing targets a maintenance step in the silencing pathway. Proc. Nat. Acad. Sci. USA 97, 13401-13406.

49. Staub, J.M., Garcia, B., Graves, J., *et al.*, High yield production of a human therapeutic protein in tobacco chloroplasts. (2000) Nat Biotechnol., 18, 330-338

50. Hood, E.E., Kusnadi, A., Nikolov, Z.L., Howard, J.A. (1999) Molecular farming of industrial proteins from transgenic maize. Adv. Exp. Med. Biol., 464, 127-147.

51. Parmenter, D.L., Boothe, J.G., van Rooijen. G.J., Yeung, E.C., Moloney, M.M. (1995) Production of biologically active hirudin in plant seeds using oleosin portioning Plant Mol. Biol., 29, 1167-1180.

52. Horvath, H., Huang, J., Wong, O., Kohl, E., Okita, T., Kannagara, C. G., von Wettstein, D. (2000) The production of recombinant proteins in transgenic barley grains Proc. Nat. Acad. Sc., USA 97, 1914-1918.

53. Wurm, F and Bernard, A. 1999 Large scale transient expression in mammalian cells for recombinant protein production. Curr. Top. Biotechnol., 10, 156-159

54. Cachianes, G., Ho, C., Weber, R.F., Williams, S.R. and Goeddel D.V. (1993) Epstein-Barr virus derived vectors for transient and stable expression of recombinant proteins. Biotechniques 15, 255-259

55. Gorman, C.M., Giles, D.R. and McCray, G. (1990) Transient production of proteins using adenovirus transformed cell line. DNA Protein Eng. Techniques 2, 1-28

56. Kim, D.W., Uetsuki, T., Kaziro, Y, Yamaguchi, N and Sugano, S. (1990) Gene 91, 217-223

57. Zhou, W, Chen, C., Buckland, B. and Aunins. J. (1997) Fed Batch Culture of recombinant NS0 myeloma cells with high monoclonal antibody production. Biotechnol. Bioengg., 55, 783-792

58. Lundstorm, P. (1997) Alphavirueses as expression vectors Curr. Opin Biotechnol., 5, 495-500

59. Blasey, H.D., Lundstrom. K., Tate. S., and Bernard. A.R. (1997) Recombinant proteins produced by Semliki Forest Virus expression system Cytotechnology, 24, 65-72

60. Paul. N.L., Marsh. M., McKeating. J.A., Schultz, T.F., Lilejstrom, P., Garoff, H. and Weiss, R.A. (1993) Expression of HIV-1 envelope glycoproteins by Semliki Forest virus vectors AIDS Res Hum Retroviruses 9, 963-970

61. Blasey. H.D., Brethon. B., Hovius. R., Lundstorm. K., Rey. L., Tairi. A.P. and Bernarad. A.R. In New Developments and New Applications in Animal Cell Technology. Merten. O.W., Perrin. P., Griffiths. P, Dordrecht, 1998, Kluwer Academic Publishers, The Netherlands, pp 449 455

62. Massie. B, Dionne. J., lamarche. N., Fleurent. J., and Langelier. Y. (1995) Improved adenovirus provides herpes simplex virus ribonucleotide reductase R1 and R2 subunits very efficiently Bio/ Technology 13, 602-608

63. Garnier. A., Cote. J., Nadeau. I., Massie. B., Chavarie. C., and Kamen. A. (1994) Scale up of the adenovirus expression system for the production of recombinant protein in human 293 cells Cytotechnology 15, 145-155

64. Pialoux. G., *et al.* (1995) A prime-boost approach to HIV preventive vaccine using a recombinant canarypox virus expressing glycoprotein 160 (MN) followed by a recombinant glycoprotein 160 (MN/LAI) AIDS Res Hum Retroviruses 11, 373-381

65. Jenkins. N, Parekh, R.B. and James D.C. (1996) Getting the glysosylation right: implications for the biotechnology industry Nature Biotechnol. 14, 975-981

66. Grabenhorst. E., Schlenke. P, pohl., S., Nimtz. M. and Conradt. H.S. (1999) Genetic engineering of recombinant glycoproteins and the glycosylation pathway in mammalian host cells Glycoconjugate 16, 81-97

67. Weikert.S. *et al.* (1999) Engineering chinese hamster ovary cells to maximize sialic acid content of recombinant glycoprteios Nature Biotechnol. 17, 1116-1121

68. Chen. K, Liu. Q, xie, L., Sharp. P.A. and Wang. D.I.C. (2001) Engineering of a mammalian cell line for reduction of lactate formation and high monoclonal antibody production Biotechnol. Bioengg. 72, 55-61

69. Laken. H.A., Leonard. M.W. (2001) Understanding and modulating apoptosis in industrial cell culture Curr. Opin. Biotechnol. 12, 175-179

70. Kim. N.S. and Lin, G.M. (2001) Overexpression of Bcl-2 inhibits sodium butyrate induced apoptosis in chinese hamster ovary cells resulting in enhanced humanized antibody production Biotechnol. Bioengg. 73, 211-222

Advances in Biochemistry and Biotechnology Volume 1 *Pages 122-157*
Edited by: **Dr. Chiranjib Chakraborty**
Published by: **DAYA PUBLISHING HOUSE**

6

Programmed Cell Death: An Overview

Satyendra Gautam & Arun Sharma*
**Head, Food Technology Division, B.A.R.C., Mumbai- 400 085,*
Email: ksarun@apsara.barc.ernet.in

ABSTRACT

Programmed cell death (PCD) also termed, apoptosis is a genetically encoded cell death program. It ensures equilibrium between cell proliferation and cell death by eliminating damaged or unwanted cells. PCD has been observed in a widely diversified group of organisms. These include prokaryotes such as bacteria, lower eukaryotes like yeasts, *Caenorhabiditis elegans*, *Drosophila melanogaster*, higher organisms, and mammals, including humans. In a bacterial population quorum sensing, or loss of plasmid may lead to PCD. In eukaryotes PCD has been observed to take place during embryonic development, metamorphosis and homeostasis. Uniqueness of the PCD is demonstrated in the conserved nature of molecular and biochemical machinery involved. A caspase-3 like protein from a bacterium (*Xanthomonas*) has been found to have homology with human caspase. Ced-3, ced-4, and ced-9 genes of *C. elegans* have been reported in mammals as homologues of caspase, Apafs (apoptotic protease activating factors), and bcl-2 genes, respectively.

* Correspondent Author